Examining Biological Foundations of Human Behavior

Barre Vijaya Prasad
Dharwad Institute of Mental Health and Neurosciences, India

A volume in the Advances in
Medical Diagnosis, Treatment, and
Care (AMDTC) Book Series

IGI Global
DISSEMINATOR OF KNOWLEDGE

Published in the United States of America by
IGI Global
Medical Information Science Reference (an imprint of IGI Global)
701 E. Chocolate Avenue
Hershey PA, USA 17033
Tel: 717-533-8845
Fax: 717-533-8661
E-mail: cust@igi-global.com
Web site: http://www.igi-global.com

Library of Congress Cataloging-in-Publication Data

Names: Vijaya Prasad, Barre, 1981- editor.
Title: Examining biological foundations of human behavior / Barre Vijaya Prasad, editor.
Description: Hershey, PA : Medical Information Science Reference, [2020] | Includes bibliographical references and index. | Summary: "This book examines the latest studies and research in the fields of biopsychology and neuropsychology"--Provided by publisher.
Identifiers: LCCN 2019044369 (print) | LCCN 2019044370 (ebook) | ISBN 9781799828600 (hardcover) | ISBN 9781799828617 (ebook)
Subjects: MESH: Neuropsychology--methods | Brain--physiology | Behavior--physiology | Brain Mapping
Classification: LCC QP360 (print) | LCC QP360 (ebook) | NLM WL 103.5 | DDC 612.8--dc23
LC record available at https://lccn.loc.gov/2019044369
LC ebook record available at https://lccn.loc.gov/2019044370

This book is published in the IGI Global book series Advances in Medical Diagnosis, Treatment, and Care (AMDTC) (ISSN: 2475-6628; eISSN: 2475-6636)

British Cataloguing in Publication Data
A Cataloguing in Publication record for this book is available from the British Library.

All work contributed to this book is new, previously-unpublished material.
The views expressed in this book are those of the authors, but not necessarily of the publisher.

For electronic access to this publication, please contact: eresources@igi-global.com.

Advances in Medical Diagnosis, Treatment, and Care (AMDTC) Book Series

ISSN:2475-6628
EISSN:2475-6636

MISSION

Advancements in medicine have prolonged the life expectancy of individuals all over the world. Once life-threatening conditions have become significantly easier to treat and even cure in many cases. Continued research in the medical field will further improve the quality of life, longevity, and wellbeing of individuals.

The **Advances in Medical Diagnosis, Treatment, and Care (AMDTC)** book series seeks to highlight publications on innovative treatment methodologies, diagnosis tools and techniques, and best practices for patient care. Comprised of comprehensive resources aimed to assist professionals in the medical field apply the latest innovations in the identification and management of medical conditions as well as patient care and interaction, the books within the AMDTC series are relevant to the research and practical needs of medical practitioners, researchers, students, and hospital administrators.

COVERAGE

- Critical Care
- Internal Medicine
- Medical Procedures
- Patient-Centered Care
- Experimental Medicine
- Disease Prevention
- Emergency Medicine
- Medical Testing
- Diagnostic Medicine
- Chronic Conditions

IGI Global is currently accepting manuscripts for publication within this series. To submit a proposal for a volume in this series, please contact our Acquisition Editors at Acquisitions@igi-global.com or visit: http://www.igi-global.com/publish/.

The Advances in Medical Diagnosis, Treatment, and Care (AMDTC) Book Series (ISSN 2475-6628) is published by IGI Global, 701 E. Chocolate Avenue, Hershey, PA 17033-1240, USA, www.igi-global.com. This series is composed of titles available for purchase individually; each title is edited to be contextually exclusive from any other title within the series. For pricing and ordering information please visit http://www.igi-global.com/book-series/advances-medical-diagnosis-treatment-care/129618. Postmaster: Send all address changes to above address.

Titles in this Series

For a list of additional titles in this series, please visit:
https://www.igi-global.com/book-series/advances-medical-diagnosis-treatment-care/129618

Multidimensional Perspectives and Global Analysis of Universal Health Coverage
Yeter Demir Uslu (Istanbul Medipol University, Turkey) Hasan Dinçer (Istanbul Medipol University, Turkey) and Serhat Yüksel (İstanbul Medipol University Turkey)
Medical Information Science Reference • © 2020 • 499pp • H/C (ISBN: 9781799823292) • US $265.00

Handbook of Research on Medical Interpreting
Izabel E.T. de V. Souza (Osaka University, Japan) and Effrossyni (Effie) Fragkou (National and Kapodistrian University of Athens Greece)
Medical Information Science Reference • © 2020 • 511pp • H/C (ISBN: 9781522593089) • US $345.00

Quality Assurance in the Era of Individualized Medicine
Anastasius S. Moumtzoglou (P. & A. Kyriakou Children's Hospital, Greece)
Medical Information Science Reference • © 2020 • 346pp • H/C (ISBN: 9781799823902) • US $285.00

Diagnosing and Managing Hashimoto's Disease Emerging Research and Opportunities
Mahmoud Sakr (Alexandria University, Egypt)
Medical Information Science Reference • © 2020 • 263pp • H/C (ISBN: 9781522596554) • US $245.00

Using Narrative Writing to Enhance Healing
Jennifer Lynne Bird (Oxbridge Academy, USA)
Medical Information Science Reference • © 2020 • 355pp • H/C (ISBN: 9781799819318) • US $275.00

Strategies to Overcome Superbug Invasions Emerging Research and Opportunities
Dimple Sethi Chopra (Punjabi University, India)

For an entire list of titles in this series, please visit:
https://www.igi-global.com/book-series/advances-medical-diagnosis-treatment-care/129618

IGI Global
DISSEMINATOR OF KNOWLEDGE

701 East Chocolate Avenue, Hershey, PA 17033, USA
Tel: 717-533-8845 x100 • Fax: 717-533-8661
E-Mail: cust@igi-global.com • www.igi-global.com

Table of Contents

Detailed Table of Contents

Neurons are the elementary structures that process information in the CNS. They are electrically excitable cells that process and transmit information around the nervous system. Neurons communicate information either by electrical or by chemical signaling that occurs through synapses. Neurons are the main part of nervous system consisting of a number of specialised types. Neuronal structure is similar to other body cells, having common features: Neurons are enclosed in a cell membrane known as a plasma membrane. The nucleus of a neuron compromises of chromosomes and genetic information. Neurons comprise of cytoplasm, mitochondria, and other organelles. The cellular processes take place in a neuron. Ribosomes are used for production of proteins, and mitochondria are for metabolic action acting as powerhouse of cell. Neurons comprise a Golgi complex that has a system of vesicles that secretes hormones and other products. The difference in neurons or nerve cells from other body cells is that extensions originate from the central body of the neuron.

Biochemistry is the cornerstone of biological science, feeding the new discoveries and providing us essential and fundamental discoveries vital to understand natural science. Biochemistry has been long recognized as a central pillar in neurochemistry, especially the imbalance of the enzyme that leads to the faulty functioning of the brain. Furthermore, we had also comprehensively discussed the metabolic and genetic aspect of major mental disorders and the genetics of intellectual disability and behavioral disorders.

Control of motor function—our activities like walking, lifting an object, writing, etc.—is accomplished through integrated and coordinated action of motor neurons. Motor neurons can be classified into two broad categories: upper motor neurons and lower motor neurons. Upper motor neurons have cell bodies in the motor cortex of brain and carry the impulses from cortex to spinal cord segment. Cell bodies of lower motor neurons are located in the anterior horn of spinal cord. Axons of lower motor neuron end in neuromuscular junction, and excitation of muscle fibres leads contraction of muscle. Different types of receptors carry different sensations like touch, pain, position, and vibration through sensory nerve fibres. Fibers mediating fine touch and proprioception ascend in the dorsal columns to the medulla, where they synapse in the gracile and cuneate nuclei. This chapter explores the neurobiology of sensory motor systems and the internal environment.

Neurotransmitters play a major role in everyday life and functioning. Everything known about human behaviour suggests it is regulated entirely by the human brain. Brain cells (neurons) communicate with one another (synaptic transmission) and with other cells in the body through small molecules called neurotransmitters (NT). NT are released by neurons and picked up by targeted cells through NT receptors (NTR). Increase or decrease in the production of any of these molecules due to any reason can produce profound effects on behaviour. Knowledge of the pathways involved in NT function has allowed development of drugs that modulate these pathways up or down. Scientists do not yet know exactly how many neurotransmitters exist, but more than 200 chemical messengers have been identified.

Neuropsychology is one of the branches in neuroscience (or sub-branch of psychology) that seeks to understand the brain structure and further relates it with specific human behavior and psychological processes. Neuropsychology's central aim is to understand how the brain produces behavior, mental processes, and cognition and then applies it for diagnosis and treatment of different behavioral and cognitive defects in various neurological disorders. Neuropsychology is a field that draws information and material from neurology, cognitive psychology, and psychiatry and also act as a junction point for investigators belonging to related disciplines. Its central aim is to understand how brain functioning influences behavior and cognition. Further,

In the current chapter, the neuropsychological profile of various neurological and psychiatric conditions is focused on, including two major divisions (i.e., dementia—cortical and sub-cortical—and major mental disorders—substance use disorders, mood disorders, anxiety disorders, psychotic disorders, sleep disorders, childhood disorders, personality disorders, and sexual disorders). Both divisions have sub-classifications that include introduction, etiopathogenesis, epidemiology, clinical features, evaluation, treatment, and psychosocial aspects.

Neuroimaging plays a crucial role in psychiatry and mental health as it can potentially be used to identify causal factors, biomarkers of a disorder, prognosis or treatment, elucidate biological pathways along with informing and monitoring newer therapies. Early and prophylactic interventions present an emerging future direction in clinical psychiatry, and neuroimaging has the potential to help in identification of individuals at risk and monitor the effects of intervention. In the chapter an attempt is made to elaborate the concept of brain mapping, and differences between structural and functional brain imaging are discussed. The chapter focusses on advent of neuroimaging in psychiatry along with discussing the major functional neuroimaging techniques.

This chapter will cover the basic concepts of neuropsychological assessment. This chapter will cover the main issues involved while assessing the person with brain damage. The process of neuropsychological assessment, relevant history for the purpose of comprehensive reporting of findings and responding to the referral questions, better understating of the goals of assessment, selection of the assessment approaches, issues involved in administration of the tests, theories involved while interpreting the test findings, and reporting of findings of the assessment will help the reader to understand holistic view of the individual's capacities.

Foreword

The book *Examining Biological Foundations of Human Behavior* edited by Dr Vijay Prasad Barre is an important addition to the existing literature on the subject. Human behavior is very complex, regulated by one of the most complex and advanced organ of the human being. Human behaviour was conventionally thought to be determined by the psychological, sociological and environmental factors, though brain has often been accepted as the prime seat determining the behavior. Scientists, psychologists, behaviourists, social scientists and neurologists have attempted to understand the determinants of the human behaviour with Sigmund Freud, Ivan Pavlov, BF Skinner and Jean Piaget being a few of them. In the last few decades, with the advances in structural and functional imaging, it has been possible to study the structure and function of the living brain, giving more insights into determinants of human behavior.

Examining Biological Foundations of Human Behavior examines the biological basis of human behavior in the background of recent advances in neurosciences, including understanding of neuroanatomy, neurophysiology, neurochemistry and neuropsychology. This book has a target readership of professionals and researchers working in the field of mental health including psychiatrists, neurologists, neurosurgeons, psychologists, social workers, psychiatric nursing, psychosocial rehabilitation therapists, general physicians and social scientists. It aims and focuses to provide with an individual's experience and innovative research and practice in the field of Mental Health.

The book has 13 chapters, grouped into two main sections. Section one has six chapters that are focused on providing in-depth information on biopsychology for example: introduction to biopsychology, anatomy of the brain, structure and function of cells, biochemistry of the brain, Neurobiology of sensory-motor systems and internal environment and neurotransmitters: behaviour. Section two contains seven chapters that were focused on Neuropsychology. It covered foundation of neuropsychology, temporal lobe syndrome, Parietal and occipital lobe syndromes, neuropsychological profile of various neurological and psychiatric conditions, functional human brain mapping: QEEG, EP & ERP, PET, SPECT, fMRI, neuropsychological assessment, and neuropsychological rehabilitation.

Dr. Barre has been in profession for more than 15 years and has been able to get a number of experienced professionals from psychology, psychiatry and neurosciences to contribute to this book. Hence this book shall emerge as a precious asset to people in desire to focus on Biological Foundation of Human Behaviour. The book has been meticulously divided into two sections covering Biopsychology and Neuropsychology together.

I hope that this book will serve its purpose by contributing in the field and will of great help to all health professionals specifically to those from psychology, mental health, neurosciences and social sciences.

I wish all the best to Dr Vijay Prasad Barre and all the authors contributing to this book.

Rakesh K. Chadda
Department of Psychiatry, All India Institute of Medical Sciences, India &
National Drug Dependence Treatment Centre, India

Preface

INTRODUCTION

Biopsychology is a branch of psychology that analyses how the brain and neurotransmitters influence our behaviors, thoughts and feelings. It is a subdivision of behavioral neuroscience that studies the neural mechanisms of perception and behavior through direct manipulation of the brains of nonhuman animal subjects in controlled experiments. Biopsychology studies many topics relating to the body's response to a behavior or activity in an organism. It concerns the brain cells, structures, components, and chemical interactions that are involved in order to produce actions. Psychologists in this field usually focus their attention to how the nervous system intertwines with other systems in the body to create a specific behavior.

Neuropsychology is an experimental field of psychology that aims to understand how behavior and cognition are influenced by brain functioning and is concerned with the diagnosis and treatment of behavioral and cognitive effects of neurological disorders. Clinical neuropsychology is the application of neuropsychological knowledge to the assessment, management, and rehabilitation of people who have suffered illness or injury (particularly to the brain and various lobe syndromes) which has caused neurocognitive problems. In particular they bring a psychological viewpoint to treatment and to understand how such illness and injury may affect and be affected by psychological factors. Technology for studying the nervous system and brain has tremendously advanced with access to tools such as PET/SPECT and MRI scans making the biological perspective in psychology increasingly important.

Brain disorders sometimes cause symptoms that look remarkably like other functional psychological disorders. The current researches have also indicated that many pharmacological agents dramatically alter the severity and course of certain mental disorders, particularly the more severe ones. This field of psychology takes an empirical and practical approach when studying the brain and human behavior. Most scientists in this field believe that the mind is a phenomenon that stems from the nervous system. Unlike other subdivisions within biological psychology, the main focus of psychological research is the development of theories that describe brain-behavior relationships. Learning how brain is involved in the genesis of normal and abnormal behavioral/emotional manifestation would result in better clinical judgment, lesser diagnostic errors and increase sensitivity to consider and rule out a neuropsychological origin or biochemical mediation of the psychopathology. The aim of this book was to provide important biological foundations of human behavior and various neurological syndromes. The main focus was the nervous system and its command centre – the brain.

This book incorporates the latest studies and research in the fields of biopsychology and neuropsychology. The theme of chapters and subtopics were helps readers to apply this information for their examination and daily routine practices. *Examining Biological Foundations of Human Behavior* is an essential text book for the professors and instructor who wants understand and teach their students in a better way. The main intension of this project to fulfill the students/instructor needs from a variety of disciplines, including clinical psychology, neurology and cognitive neuroscience.

OBJECTIVE OF THE BOOK

Thirteen chapters were included. We asked researchers to submit chapters and look at various biological factors, structure of the brain and behavior and intervention or treatment issues as well. It will also integrate the findings from the individual studies if any. We believe that research chapters in the collection will make significant contributions to a currently under-researched aspect of neuroscience, and the collection as a whole were set an agenda at global level to guide ongoing re conceptualizations of the individual's experience and innovative research, practice and giving importance in the field of Neuroscience.

TARGET READERS

The target readers of this book were professionals and researchers working in the field of mental health: neuropsychologists and clinical psychologists, neuroscience neurology, neurosurgeons, psychiatrists, including psychiatric nursing, psychosocial rehabilitation therapists, mental health counselors, general physicians, and social scientists. The book can also serve as a complementary textbook for advanced courses in the area of Neurobiology, medical sciences of various diseases, multidisciplinary courses designed to undergraduates, postgraduates and doctorate research students in the field of mental health. This book will be a referral book for academic purposes in various libraries.

ORGANIZATION OF THE BOOK

There are 13 chapters in this book and the chapters are grouped into two main sections. The two sections assist the reader to be guided to a specific area of interest because each section is based on a particular theme. The sections are divided as follows: Section 1 is biopsychology and contains six chapters that are focused on providing in-depth information anatomical and physiological aspects of the brain. Section 2 contains seven chapters and was focused on neuropsychological basis of behaviour emphasizing on various lobes and its functions, neuropsychological profiles of psychiatric conditions, neuroimaging methods, neuropsychological assessments and neuro-rehabilitation training concern. All these sections were summarized briefly in following order.

BIOPSYCHOLOGY

Chapter 1: Basics of Biopsychology

The author emphasized the subject of biopsychology and to describe the physiological mechanisms of the body that mediate our movement and mental activity. Biopsychology focuses on biological basis of behavior i.e. how brain and other biological processes affect psychological behaviors. Biopsychology is also known as biological psychology or psychobiology. Further, this chapter was covered under following subheadings such as scope of biopsychology covered the six subdivisions i.e., Physiological psychology, psychopharmacology, neuropsychology, psychophysiology, cognitive neuroscience and Comparative psychology. In addition, the historical and scientific roots of biopsychology, the mind-body interactions, neurobiology and its philosophy, biopsychological research methods and lesions methods discussed

Chapter 2: Anatomy of the Brain

Brain is the most complex structure present in the human body. It is responsible for the thoughts, memory, and speech, different bodily movements and also regulates the movement and function of internal organs present inside the body. Hence brain anatomy is crucial to understand human psychology and also develop suitable treatments to re-correct it or even remove glioma. Brain is a floating structure in cerebrospinal fluid covered by the hard bony skull, and then three thin membranes called meninges. It is divided into cerebrum, brainstem, and cerebellum. Among all cerebrum is the biggest part present in the brain and further divided into two regions, namely left and right cerebral hemispheres. Cell staining, histochemistry, and non-invasive brain imaging are the major tools used for the study of brain anatomy were covered by the author.

Chapter 3: Structure and Function of Cell

Neurons are the elementary structures which process information in the Central Nerves System (CNS). They are electrically excitable cells that can receive and transmit chemical or electrical signals rapidly through synapses around the nervous system. Neurons are the main part of nervous system consisting of number of specialized type and function. Neuronal structure is similar to other body cells, having similar features, yet there are complexities vary throughout animal kingdom. It change from nerve nets (neurons) in lower organism to centralized structure consisting central nervous system (CNS) made of brain, spinal cord and a peripheral nervous system (PNS) made up of peripheral sensory and motor nerves. Further this chapter is illustrated with neuron structure, neuron and its characteristics, neuron classification, neuroglial cells, the action potential and nerve impulse, synaptic transmission, the synaptic vesicle, modulation of synaptic transmission, non-synaptic chemical communication, postsynaptic receptors and receptor types, neurotransmitters, cell structure, neuropeptides and neuromodulators are covered with suitable diagrams. The current chapter authors were highlighted about the Neurons and its function of cell without difficulty.

Chapter 4: The Biochemistry of the Brain

In this chapter authors were highlighted the biochemistry is the cornerstone in the biological science, feeding the new discoveries and providing us essential and fundamental discoveries vital to understand natural science. Biochemistry has been long recognized as central pillar in neurochemistry, especially the imbalance of the enzyme which, leads to the faulty functioning of the brain. Biochemical and metabolic aspect of major mental disorders are always traced to some metabolic deficiencies. These deficiencies linked with four hundred types of mental disorders are identified and placed into five major categories are as follows: disorders diagnosed during childhood, organic mental disorders, substance-related disorders, mood disorders and personality disorders, schizophrenia and other psychotic disorders. This chapter is also focused the genetics of mental retardation (Intellectual disability). Other genetic disorders such as dominant genes disorder, recessive genes disorder, phenylketonuria, sex-linked Inheritance in mental retardation, lesch-Nyhan Disease, Fragile X Syndrome, Down syndrome, Prader-Willi Syndrome, Williams Syndrome, Klinefelter syndrome, Turner syndrome. In addition to other behavioural disorders covered such as generalized anxiety disorder, substance abuse disorder, eating disorder, learning disorder, conduct disorder and comorbidity of behavioral disorders.

Chapter 5: Neurobiology of Sensory-Motor Systems and Internal Environment

The brain is not an independent agent, residing in splendid and lofty superiority in our skulls. Rather it is part of an extended system reaching out to permeate, influence, and be influenced by, every corner and extremity of your body. As the spinal cord, our brain extends the length of the backbone, periodically sprouting nerves that convey information to and from every part of our body. This chapter explains about physiology of the brain and clinical implications of sensory system, motor system and regulation of internal environment and its sub topics limbic system, autonomic system, neuroendocrine system and discussed briefly by author in this chapter to understand by far with pictorial presentations.

Chapter 6: Neurotransmitters and Behavior

This chapter was highlights by the author the role of relationship of neurotransmitters with behavior. It includes specifically neurotransmitters classification according to its chemical structure and mechanism of actions, its types etc. Neurotransmitters mechanism of action in various mental disorders is discussed in details to help reader develop the understanding while dealing such conditions in their practice. Followed by major neurotransmitters, disorders associated with neurotransmitters systems and role of serotonin, glutamate, gcetylcholine described. Further this chapter also explained as subheads neurotransmitter basis of learning, neurotransmitters adverse impacts, neurotransmitter dysfunction in behavioural disorder, inhibitory Neurotransmitters, neurotransmitters influenced by the various drugs explained in precisely.

NEUROPSYCHOLOGY

Chapter 7: Foundation of Neuropsychology

This chapter was largely covered by authors in introduction as foundation of neuropsychology is a field that draws information and material from neurology, cognitive psychology, and psychiatry and also act as a junction point for investigators belongs to related discipline discussed. Further, in relationship between structure and function of the brain, the rise of neuropsychology as a distinct discipline, origin of brain hypothesis, and logic of cerebral organization covered. Localization and lateralization of brain function: in these headings certain functions for example language and memory, and functions of lateralization were explained. Further, approaches and methodologies of clinical and cognitive neuropsycholoigsts and methods such as standardized neuropsychological tests are illustrated with suitable images.

Chapter 8: Temporal Lobe Syndrome

The brain's cerebral cortex is the outermost part of the brain which contains two cerebral hemispheres connected by the corpus callosum. Traditionally, each of these cerebral hemispheres is divided into four lobes: Frontal, parietal, temporal and occipital. Further, subtopics were covered by authors in details viz., Anatomy of temporal lobe, functional anatomy of temporal lobe, connections of the temporal lobe, temporal lobe syndromes. More specifically under temporal lobe syndromes other issues were also illustrated such as Kluver-Bucy syndrome, Geschwind-Gastaut syndrome, Wernicke's aphasia, Cortical deafness, Delusional misidentification, Auditory paracusias, auditory agnosia, amusia, prosopagnosia. Following neuropsychology of the temporal lobe: Assessment and its implications discussed with conclusion.

9: Parietal and Occipital Lobe Syndromes

Human brain is the most important as well as the most complex organs in the human body. From previous chapters, we by now know that the cerebral cortex is the part of the brain that functions to make human beings unique. Distinctly human traits including higher thought, language, and human consciousness as well as the ability to think, reason and imagine all originate in the cerebral cortex. The cerebral cortex is the outermost portion that can be divided into the four lobes of the brain. Further this chapter was covered by authors under following headings neuroanatomy of parietal lobe, functioning of parietal lobe, parietal lobe dysfunction, neuroanatomy of occipital lobe, functioning of occipital lobe and occipital lobe dysfunction.

Chapter10: Neuropsychological Profile of Various Neurological and Psychiatric Conditions

Dementia is a progressive impairment in the functions of the brain especially of the cognitive domain when there is clear conscious. This leads to decline in the previously learned things in more than one domains executive function, attention, language, memory, complex motor tasks, following social norms, these will lead to decline in the level of functioning. This chapter explicate by the author about the neuropsychological profile of cortical and subcortical dementia under following headings, etiopathogenesis, epidemiology, clinical features, evaluation, treatment, psychosocial aspects. Further, major mental disorders were described under the following headings such as substance use disorders, mood disorders, anxiety disorders, psychotic disorders, sleep disorders, childhood disorders, personality disorders and sexual disorders, were broadly covered to understand subject concern.

Chapter 11: Functional Human Brain Mapping

Neuroimaging or brain imaging is the use of various techniques to either directly or indirectly image the structure, function, or pharmacology of the nervous system. It is a relatively new discipline within medicine, neuroscience and psychology. Neuroimaging is the discipline that deals with the *in vivo* depiction of anatomy and function of the central nervous system (CNS) in health and disease. This chapter largely focuses by author in following headings, history of neuroimaging and brain mapping, difference between structural and functional brain mapping, importance of brain mapping in psychiatric disorders and methods of study in brain mapping.

Chapter 12: Neuropsychological Assessment

Neuropsychological assessment refers to the measurement of cognitive functions and processes with the aim to establish whether cognitive dysfunction or cognitive impairment is present in individuals, typically in patients with (suspected) brain disease, psychiatric disorder or information-processing complaints. Neuropsychological assessment is preferably performed by clinical neuropsychologists who have expertise in the study of the neural correlates of behavior and cognition. This chapter covered by the author with sub topics such as history of neuropsychological assessment, types of neuropsychological assessment using in India, goals of neuropsychological Assessment, approaches of neuropsychological assessment, neuropsychological test findings interpretation, and neuropsychological assessment process.

Chapter 13: Neuropsychological Rehabilitation

Neuropsychological rehabilitation is concerned with achievement of maximum potential in diverse domains of psychological, social, vocational and everyday life functioning in people with cognitive, emotional, psychosocial, and behavioral deficits caused by an insult to the brain. Hence neuropsychological rehabilitation is the process of helping a patient to recover the functions which are impaired due to brain injury. Further, the authors were focuses on Neuropsychological rehabilitation how is used for people with various conditions such as, traumatic brain Injury (TBI), stroke, tumor, epilepsy, degenerative conditions, Alzheimer's disease, schizophrenia, attention-deficit disorder, learning disabilities and interventions methods were highlighted in brief.

Barre Vijay Prasad
Dharwad Institute of Mental Health and Neurosciences, India

REFERENCES

Beaumont, J. G. (1983). *Introduction to neuropsychology*. Oxford, UK: Blackwell Scientific Publications.

Carlson, N. R. (2008). *Foundations of physiological psychology*. Boston, MA: Allyn and Bacon.

Kolb, B., & Whishaw, I. Q. (2009). *Fundamentals of human neuropsychology*. New York, NY: Worth Publishers.

Tyerman, A., & King, N. S. (2013). Interventions for psychological problems after brain injury. In L. H. Goldstein & J. E. McNeil (Eds.), *Clinical neuropsychology: A practical guide to assessment and management for clinicians* (pp. 527–548). Wiley-Blackwell.

Zillmer, E., Spiers, M., & Culbertson, W. C. (2008). Principles of neuropsychology (Recent Edition). Thomson/Wadsworth.

Acknowledgment

Having an idea and turning it into a book is as hard as it sounds. The experience is both internally challenging and rewarding. I especially want to thank the all my authors who helped make this happen. I felt it's my responsibility do some piece of contribution in the field of mental health, in this journey I always remember and owe my sincere thanks to Prof. Singh.T.B. Who is like a God father, he stood by me during my professional journey and all my achievements. Secondly my mentor Prof. Padmaja, her motivation, immense knowledge and guidance helped me in all my technical writings and it's my honor to acknowledge her.

My sincere thanks goes to Prof. Srinivas Kosgi, who always available for helping me during the seminal and completion of the work and I consider him as an advisor for this book. I heartily thank my senior faculty colleague Gauri Shankar Kaloiya's for his guidance and suggestions during this project.

I would not forget to remember faculty from DIMHANS, Manjunath R. Bajantri, Gayatri Hegde and from AIIMS, Koushik S. Deb and Sujatha Satpathy for their encouragement. It's my pleasure to mention my sincere thanks to many of my senior co- faculty from AIIMS-New Delhi and NDDTC.

My special honor and worth to mention Prof and Head, Rakhesh K Chadda he is a teacher for many generations to psychiatry, and I'm very fortunate to travel with his professional headship.

Writing a book is harder than I thought and more rewarding than I could have ever imagined. None of this would have been possible without mentioning Sudarshini S. Reddy. She was the first person suggest me to write this book for Clinical Psychology Scholars from RCI and thanks for the great suggestion.

I am indebted to thanks to Sumit Kumar (JNU), who is technically devoted many hours of his attention during the process of proof reading and creating images for this manuscript.

Acknowledgment

Thank you doesn't seem sufficient but it is said thank you parents for your support and unconditional love and words would never say how grateful I am to both of you.

"Thanks to everyone of my IGI publishing team."

Barre Vijay Prasad
Dharwad Institute of Mental Health and Neurosciences, India

Chapter 1
Biopsychology:
An Introduction

Barre Vijaya Prasad
https://orcid.org/0000-0003-2332-8220
All India Institute of Medical Sciences (AIIMS), India & Dharwad Institute of Mental Health and Neurosciences, India

ABSTRACT

The study of biopsychology seeks to describe the physiological mechanisms of the body that mediate our movement and mental activity. Biopsychology focuses on biological basis of behavior (i.e., how brain and other biological processes affect psychological behaviors). Biopsychology is also known as biological psychology or psychobiology. Biopsychology is the application of the principles of biology to the study of mental processes and behavior that is the study of psychology in terms of bodily mechanisms.

CHAPTER OUTLINE

- The origins of biopsychology
- Nature and scope of biological psychology
- Mind-body interactions
- Biopsychological research methods

DOI: 10.4018/978-1-7998-2860-0.ch001

INTRODUCTION

The study of biopsychology seeks to describe the physiological mechanisms of the body that mediate our movement and mental activity. Biopsychology focuses on biological basis of behavior i.e. how brain and other biological processes affect psychological behaviors. Biopsychology is also known as biological psychology or psychobiology. Biopsychology is the application of the principles of biology to the study of mental processes and behavior that is the study of psychology in terms of bodily mechanisms. The view that psychological processes have biological (or physiological) correlates is the basic assumption of the whole field of biological psychology. Through a variety of research methods, psychologists in this field hope to uncover information that enriches human understanding of their own mental processes, as well as providing valuable data that enable those in medical fields to better treat patients with a variety of disorders, both physical and mental. Biopsychology has been a prominent field of psychology from the start in Europe and North America and remains a major area of research and instruction in many countries. In the last two centuries, biopsychology has found new ways to answer old questions, has tackled important new questions, and has abandoned some problems as poorly defined. Carefully designed behavioral experiments and innovative biomedical techniques have been essential to its progress.

NATURE OF BIOPSYCHOLOGY

Biopsychology is the study of how emotions, thoughts and behavior are affected by the brain, the nervous system and neurotransmitters in human and non-human animals. Biological psychology, behavioral neuroscience, psychobiology, and physiological psychology are another term often used synonymously with biological psychology. In order to understand something it is important to know the history and have a definition. Biological psychology has been dated to Avicenna (980-1037 C.E.), a physician who recognized physiological psychology in the treatment of illnesses and linked the physical and psychological together. Behavioral neuroscience, also known as biological psychology (Breedlove, Rosenzweig & Watson, 2007). Pinel (2003) defines Biopsychology (also known as psychobiology, behavioral biology, behavioural neurosciences) as "the scientific study of the biology of behavior". Carlson (2005) calls it the physiology of behavior where the brain and physiological correlates and modulators of behavior are the domain of study.

SCOPE OF BIOPSYCHOLOGY

Biopsychology is one of many contributors to neuroscience and adds the dimensions of behavior and behavioral research. The six primary divisions within biopsychology have three basic research dimensions. According to Pinel these may include human or nonhuman subjects, formal experiments or nonexperiments, and can be pure or applied.

There are six major divisions that demonstrate the diversity of biopsychology. Each one specifies a distinct area, yet seldom do they stand alone. In most situations a synergistic and harmonious blending of various combinations of these divisions is what delivers the most accurate insight into a specific situation.

The six divisions are:

Physiological Psychology: Examines behavior by examining the neural mechanism and manipulating the nervous system of animals. An example would be removing part of a hamster brain to examine how they perform on memory tests.

Psychopharmacology: Examines how drugs affect the brain and behavior. An example of this would be examining new drugs to help schizophrenic patients.

Neuropsychology: Examines brain damaged humans with the use of case studies or quasi-experimental studies. An example would be studying the effects of neural damage on behavior. This helps physicians in forming a clear diagnosis with appropriate medications.

Psychophysiology: Examines the relationship between physiological and psychological processes and activity. An example of this would be the relationship between nutrition and biological markers, such as eating fish and the improvement of overall health (Hansen, Dahl, Bakke, Froyland, & Thayer, 2010). This is an area of research that might have important implications with regard to both physical and mental health.

Cognitive Neuroscience: Examines neural connections in human cognition, mostly through brain imaging. An example of this would be neuropsychological work on patients with difficulties in speech perception and production that has been expanded to include nonhuman primates (Teufel, Ghazanfar, & Fischer, 2010).

Comparative Psychology: Compares evolution, genetics, and other adaptive behavior as well as the comparison of theories. An example of this would be Darwin's theory of evolution (Burghardt, 2009). Biopsychologists have an in-depth understanding of anatomy and physiology. When trying to ascertain the underlying causes of normal and abnormal behavior, they inquire at the level of nerves, brain circuitry, neurotransmitters and basic biological processes. Researchers often focus on how damage to specific areas of the brain affects neural function and behavior, as well as the influence of drugs and other mind altering substances on the brain and body. While human subjects are used in many Biopsychological experiments, most of the experimental data has been obtained via work with mice, rats, monkeys and other non-human primates.

Behavioral neuroscience has a strong history of contributing to the understanding of medical disorders, including those that fall into the realm of clinical psychology. Neuropsychologists are often employed as scientists to advance scientific or medical knowledge, and neuropsychology is particularly concerned with understanding brain injuries in an attempt to learn about normal psychological functioning. Neuroimaging tools, such as functional magnetic resonance imaging (fMRI) scans, are often used to observe which areas of the brain are active during particular tasks in order to help psychologists understand the link between brain and behavior.

HISTORICAL AND SCIENTIFIC ROOTS OF BIOPSYCHOLOGY

The history of biological psychology is a major part of the history of modern scientific psychology. The study of biological psychology can be dated back to Avicenna (980-1037 C.E.), a physician who in *The Canon of Medicine,* recognized *physiological psychology* in the treatment of illnesses involving emotions, and developed a system for associating changes in the *pulse* rate with inner feelings, which is seen as an anticipation of the word association test. Avicenna also gave psychological explanations for certain somatic illnesses, and he always linked the physical and psychological illnesses together. He explained that "humidity" inside the head can contribute to mood disorders, and he recognized that this occurs when the amount of "breath" changes: happiness increases the breath, which leads to increased moisture inside the brain, but if this moisture goes beyond its limits, the brain would lose control over its rationality and lead to mental disorders.

Biological psychology as a scientific discipline later emerged from a variety of scientific and philosophical traditions in the eighteenth and nineteenth centuries. In philosophy, the first issues is how to approach what is known as the " mind-body problems " namely the explanation of the relationship, if any, that obtains between minds, or mental processes, and bodily states or processes. Dualism is a family of views about the relationship between mind and physical matter. It begins with the claim that mental phenomena are, in some respects, non-physical. In Western Philosophy, some of the earliest discussions of dualist ideas are in the writings of Plato and Aristotle. Each of these maintained, but for different reasons, that human "Intelligence" (a faculty of the mind or Soul) could not be identified with, or explained in terms of, his physical body. However, the best-known version of dualism is due to Rene Descartes (expressed in his 1641, *Meditations on First Philosophy*), and holds that the mind is a non-extended, non-physical substance. Descartes was the first to clearly identify the mind with consciousness and self-awareness, and to distinguish this from the brain, which was the seat of intelligence.

The question then, is how do these separate and entirely different aspects of living beings, the mind and the body, relate? Some, like Descartes, proposed physical models to explain animal and human behavior. Descartes, for example, suggested that the pineal gland, a midline unpaired structure in the brain of many organisms, was the point of contact between mind and body. Descartes also elaborated on a theory in which the pneumatics of bodily fluids could explain reflexes and other motor behavior. This theory was inspired by moving statues in a garden in Paris.

Other philosophers also helped to give birth to psychology, also relating its subject matter to biology. This view, that psychological processes have biological (or physiological) correlates, is the basic assumption of the whole field of biological psychology. One of the earliest textbooks in the new field, *The Principles of Psychology* by William James (1890), argues that the scientific study of psychology should be grounded in an understanding of biology:

Bodily experiences, therefore, and more particularly brain-experiences, must take a place amongst those conditions of the mental life of which Psychology need take account. The spiritualist and the associationist must both be "cerebralists," to the extent at least of admitting that certain peculiarities in the way of working of their own favorite principles are explicable only by the fact that the brain laws are a codeterminant of their result. Our first conclusion, then, is that a certain amount of brain-physiology must be presupposed or included in Psychology.

William James, like many early psychologists, had considerable training in physiology. The emergence of both psychology and biological psychology as legitimate sciences can be traced from the emergence of physiology from anatomy, particularly neuroanatomy. Physiologists conducted experiments on living organisms, a practice that was distrusted by the dominant anatomists of the eighteenth and nineteenth centuries. The influential work of Claude Bernard, Charles Bell, and Wiliam Harvey helped to convince the scientific community that reliable data could be obtained from living subjects.

The term "psychobiology" has been used in a variety of contexts, but was likely first used in its modern sense by Knight Dunlap in his book, *An Outline of Psychobiology (1914)*. Although a "forgotten man" of American psychology, Dunlap also founded the journal *Psychobiology*. In the announcement of that journal, Dunlap writes that the journal will publish research "...bearing on the interconnection of mental and physiological functions," which describes the field of biological psychology even in its modern sense.

THE MIND-BODY INTERACTION

The mind–body problem is a debate concerning the relationship between thought and consciousness in the human mind, and the brain as part of the physical body. It is distinct from the question of how mind and body function chemically and physiologically, as that question presupposes an interactionist account of mind–body relations (Skirry, 2016). This question arises when mind and body are considered as distinct, based on the premise that the mind and the body are fundamentally different in nature (Skirry, 2016). Dated the problem was addressed by Rene Descartes in the 17th century, resulting in Cartesian dualism, and by pre-Aristotelian philosophers (Robert, 1996; Robinson, 2011), in Avicennian philosophy (Henrik, 2010), and in earlier Asian traditions. A variety of approaches have been proposed. Most are either dualist or monist. Dualism maintains a rigid distinction between the realms of mind and matter. Monism maintains that there is only one unifying reality, substance or essence, in terms of which everything can be explained.

Each of these categories contains numerous variants. The two main forms of dualism are substance dualism, which holds that the mind is formed of a distinct type of substance not governed by the laws of physics, and property dualism, which holds that mental properties involving conscious experience are fundamental properties, alongside the fundamental properties identified by a completed physics. The three main forms of monism are physicalism, which holds that the mind consists of matter organized in a particular way; idealism, which holds that only thought truly exists and matter is merely an illusion; and neutral monism, which holds that both mind and matter are aspects of a distinct essence that is itself identical to neither of them. Psychophysical parallelism is a third possible alternative regarding the relation between mind and body, between interaction (dualism) and one-sided action (monism)(Chisholm & Hugh, 1911). Several philosophical perspectives have been developed which reject the mind–body dichotomy. The historical materialism of Karl Marx and subsequent writers, itself a form of physicalism, held that consciousness was engendered by the material contingencies of one's environment (Marx, 1977). An explicit rejection of the dichotomy is found in French structuralism, and is a position that generally characterized post-war Continental philosophy9(Bryan, 2008).

The absence of an empirically identifiable meeting point between the non-physical mind (if there is such a thing) and its physical extension (if there is such a thing) has proven problematic to dualism, and many modern philosophers of mind maintain that the mind is not something separate from the body (Bryan, 2008). These approaches have been particularly influential in the sciences, particularly in the fields of sociobiology, computer science, evolutionary psychology, and the neurosciences (Pinel, 2009; Russell, & Norvig, 2010).

An ancient model of the mind known as the Five-Aggregate Model, described in the Buddhist teachings, explains the mind as continuously changing sense impressions and mental phenomena (Karunamuni, 2015). Considering this model, it is possible to understand that it is the constantly changing sense impressions and mental phenomena (i.e., the mind) that experiences/analyzes all external phenomena in the world as well as all internal phenomena including the body anatomy, the nervous system as well as the organ brain. This conceptualization leads to two levels of analyses: (i) analyses conducted from a third-person perspective on how the brain works, and (ii) analyzing the moment-to-moment manifestation of an individual's mind-stream (analyses conducted from a first-person perspective). Considering the latter, the manifestation of the mind-stream is described as happening in every person all the time, even in a scientist who analyses various phenomena in the world, including analyzing and hypothesizing about the organ brain (Karunamuni, 2015). In neuroscience, much has been learned about correlations between brain activity and subjective, conscious experiences. Many suggest that neuroscience will ultimately explain consciousness: "...consciousness is a biological process that will eventually be explained in terms of molecular signaling pathways used by interacting populations of nerve cells..." (Marx, 1977). However, this view has been criticized because *consciousness* has yet to be shown to be a *process* (Bryan, 2008) and the "hard problem" of relating consciousness directly to brain activity remains elusive (Kim, 1995).

Cognitive science today gets increasingly interested in the embodiment of human perception, thinking, and action. Abstract information processing models are no longer accepted as satisfactory accounts of the human mind. Interest has shifted to interactions between the material human body and its surroundings and to the way in which such interactions shape the mind. Proponents of this approach have expressed the hope that it will ultimately dissolve the Cartesian divide between the immaterial mind and the material existence of human beings (Damasio, 1994; Gallagher, 2005). A topic that seems particularly promising for providing a bridge across the mind–body cleavage is the study of bodily actions, which are neither reflexive reactions to external stimuli nor indications of mental states, which have only arbitrary relationships to the motor features of the action (e.g., pressing a button for making a choice response). The shape, timing, and effects of such actions are inseparable from their meaning. One might say that they are loaded with mental content, which cannot be appreciated other than by studying their material features. Imitation, communicative gesturing, and tool use are examples of these kinds of actions (Pinel, 2009).

NEUROBIOLOGY AND ITS PHILOSOPHY

A science of consciousness must explain the exact relationship between subjective conscious mental states and brain states formed by electrochemical interactions in the body, the so-called hard problem of consciousness (Kandel, 2007). Neurobiology studies the connection scientifically, as do neuropsychology and neuropsychiatry. *Neurophilosophy* is the interdisciplinary study of neuroscience and philosophy of mind. In this pursuit, neurophilosophers, such as Churchland (2002; 1989); Churchland (2007) and Daniel Dennett (1986;1997) have focused primarily on the body rather than the mind. In this context, neuronal correlates may be viewed as causing consciousness, where consciousness can be thought of as an undefined property that depends upon this complex, adaptive, and highly interconnected biological system (Squire, 2008). However, it's unknown if discovering and characterizing neural correlates may eventually provide a theory of consciousness that can explain the first-person experience of these "systems", and determine whether other systems of equal complexity lack such features. The massive parallelism of neural networks allows redundant populations of neurons to mediate the same or similar percepts. Nonetheless, it is assumed that every subjective state will have associated neural correlates, which can be manipulated to artificially inhibit or induce the subject's experience of that conscious state. The growing ability of neuroscientists to manipulate neurons using methods from molecular biology in combination with optical tools[31] was achieved by the development of behavioral and organic models that are amenable to large-scale genomic analysis and manipulation. Non-human analyses such as this, in combination with imaging of the human brain, have contributed to a robust and increasingly predictive theoretical framework.

BIOPSYCHOLOGICAL RESEARCH METHODS

Biopsychology research methods study the human brain. Since, beginning many people are fascinated to towards understanding the brain and behaviour, so far we know very limited. At the same time growing science and technology we never stop to do research about it. The technology now made easier to understand how the most mysterious organ in our body works. But what exactly are these methods?

Biopsychology uses unique concepts and measurements to explore the brain and nervous system. The human brain is extremely complicated, but biopsychology is helping people understand the many unknown functions of the human brain. Biopsychologists study the brain's functionality through neuro-surgery. They often study patients with extremely rare neurological diseases. Biopsychology research methods have evolved in the last decades. Although there are several biopsychology research methods such as human brain stimulation and visualization methods viz; contrast radiography (CR), computerized axial tomography scan (CT), nuclear magnetic resonance scan (NMR), positron Emission Tomography (PET), functional Magnetic resonance imaging (fMRI), Magnetoencepholography (MEG) and Transcranial magnetic stimulation (TMS), some important methods are presented in detail chapter no 11. (Functional human brain mapping) and that are briefly explained in following headings;

Contrast Radiography (CR): This technique consists of injecting a substance in the body to absorb X-rays. This way, scientists see the contrast between the compartment and the tissue around it. Cerebral angiography is a kind of contrast radiography. To do it, a contrast medium is inserted into a brain vessel. The aim is to observe the circulatory system while performing an X-ray. This technique is very useful to locate vessel injuries and brain tumors.

Computerized axial Tomography Scan (CT): Through a CT scan, experts can see the whole brain structure. During the test, the patient lies down in the middle of a big cylinder. While the patient holds still, an X-ray tube and a receptor take many separate photographs. This happens while the emitter and the receptor spin around the patient's head.

All this information is transmitted to a computer, which allows doctors to explore the brain on a horizontal plane. Usually, they do it on eight to nine horizontal brain sections. Once all explorations are combined, it's possible to make a three-dimensional representation of the brain.

Nuclear Magnetic Resonance (NMR): NMR facilitates high-resolution imaging thanks to the different waves hydrogen atoms emit when activated by radio frequency on a magnetic field. It provides a high spatial resolution and produces three-dimensional images.

Positron Emission Tomography (PET): PET provides brain activity images instead of brain structure images. To get the images, scientists inject radioactive fludeoxyglucose (FDG) in the carotid artery. Active neurons rapidly absorb FDG, which accumulates once neurons don't metabolize it anymore, then it slowly degrades. This is how one can observe which neurons are active at a given time during different activities.

Functional Magnetic Resonance Imaging (fMRI): On the other hand, MRIs offer an image of the increase in the amount of oxygen present in the brain's blood. Thus, it successfully measures brain activity. If we compare it to the PET, it actually has four advantages: doctors don't have to inject the patient with anything. It provides both functional and structural information. It offers a better spatial resolution. Provides three-dimensional images of the entire brain.

Magnetoencephalography (MEG): It measures the changes in the magnetic fields located on the scalp's surface. These changes occur due to the variations in the neuronal activity guidelines.

Transcranial magnetic stimulation (TMS). Walsh and Rothwell (2000) state that TMS "alters an area of the cortex, creating a magnetic field under the coil that goes over the cranium." TMS basically "turns off" a part of the brain temporarily, to study behavior and cognition under those circumstances.

LESION METHODS

Lesion methods focus on the *destruction of a small brain area* to see how it affects behavior.

Aspiration Lesion: This method creates a lesion in an exposed or easily accessible cortical tissue area. The doctors remove the tissue with a fine-point crystal pipette.

Radio-Frequency Lesion: It's carried out by creating small subcortical lesions. For that, an electrode channels the high-frequency current through the tissue of interest. The size and shape of the lesion depend on three factors:

Duration of the procedure.

Intensity of the current.

Configuration of the electrode's point.

Scalpel Cuts: It consists of sectioning the brain area of interest.

Cooling Lesion: This biopsychology research method, although included under lesion methods, is actually temporary and reversible. Instead of destroying tissue, *one area is cooled a bit over the freezing point.* Neurons stop emitting signals, so the cold brain area remains blocked. With this, researchers are able to see what behavior alterations are caused by those areas. Once the temperature goes back to normal, brain function is restored.

ELECTRICAL STIMULATION

Another biopsychology research method is electrical stimulation. The procedure consists of electrically stimulating a nervous system structure to get data about its functions. Usually, a bipolar electrode is used. This stimulation "shoots" neurons and alters their behavior. In general, it tends to get the opposite effect of lesion methods. For example, if it's possible to drastically reduce sleep hours with a lesion, then sleep behavior can become inconvenient and harder to control with electrical stimulation.

Lesion Methods with Electrical Recording

Intracellular Recording: This technique is carried out by introducing a microelectrode in the interior of a neuron. It records fluctuations of membrane potential.

Extracellular Unit Recording: A microelectrode is placed in the extracellular fluid that surrounds the neuron. It doesn't provide information on membrane potential.

Multi-Unit Recording: In this case, the electrode point is bigger than that of a microelectrode, so it captures the signals of many neurons at the same time. The detected potentials then go on a circuit that integrates them.

Invasive EEG Monitoring: The stainless steel electrodes go inside the cranium. For subcortical signals, the usual electrodes are made of cable and implanted through stereotactic surgery.

Alongside, we can see many more methods and techniques in coming years for research on brain and to improve human existence with quality of life and disease free inventions.

CONCLUSION

Biopsychology is an important aspect of psychology that assists in the exploration of the biological aspect of psychology. It involves the contribution of anatomical and physiological functional processes in perceptual processes. This enhances the elaboration of biopsychology as an internal body factor that is closely related to nervous system. In addition, the theories of attention and perception help in the analysis of biopsychology through the irelements. Although there are assumptions behind biopsychology, the area of study has advanced making it to be one of the common psychology topics. Therefore, biopsychology covers broad aspects of human behavior were illustrate in following chapters.

REFERENCES

Breedlove, S. M., Rosenzweig, M. R., & Watson, N. V. (2007). *Biological psychology: An introduction to behavioral, cognitive, and clinical neuroscience* (5th ed.). Sinauer Associates.

Bryan, S. T. (2008). *The Body and Society: Explorations in Social Theory* (3rd ed.). Sage Publications.

Bryan, S. T. (2008). *The Body and Society: Explorations in Social Theory* (3rd ed.). Sage Publications.

Burghardt, G. M. (2009). Ethics and Animal Consciousness: How Rubber the Ethical Ruler? *The Society for the Psychological Study of Social Issues*, *65*(3), 499–521. doi:10.1111/j.1540-4560.2009.01611.x

Carlson. (2005). *Foundations of Physiological Psychology*. Allyn and Bacon.

Chisholm, H. (Ed.). (1911). *Parallelism, Psychophysical. In Encyclopædia Britannica* (11th ed.; vol. 20, p. 762). Cambridge University Press.

Churchland, P. (2007). Neurophilosophy at Work. Cambridge University Press. doi:10.1017/CBO9780511498435

Churchland, P. S. (1989). Neurophilosophy: Toward a Unified Science of the Mind–Brain. In Computational Models of Cognition and Perception. MIT Press.

Churchland, P. S. (2002). Brain-Wise: Studies in Neurophilosophy. Bradford Books.

Damasio, A. R. (1994). *Descartes' error: emotion, reason, and the human brain*. New York: Grosset/Putnam.

Dennett, D. C. (1986). Content and Consciousness. International Library of Philosophy. Taylor & Francis.

Dennett, D. C. (1997). Kinds of Minds: Toward an Understanding of Consciousness. Science Masters Series. Basic Books.

Gallagher, S. (2005). How the Body Shapes the Mind. New York: Oxford University Press. doi:10.1093/0199271941.001.0001

Henrik, L. (2010). Introduction. In Forming the Mind: Essays on the Internal Senses and the Mind/Body Problem from Avicenna to the Medical Enlightenment. Springer Science+Business Media.

James, W. (1890). *The principles of psychology*. New York: Holt.

Kandel, E. R. (2007). *In search of memory: The emergence of a new science of mind*. W. W. Norton & Company.

Karunamuni, N. D. (2015). The Five-Aggregate Model of the Mind. *SAGE Open*, *5*(2). doi:10.1177/2158244015583860

Kim, J. (1995). Emergent properties. In T. Honderich (Ed.), *Problems in the Philosophy of Mind. Oxford Companion to Philosophy* (p. 240). Oxford, UK: Oxford University Press.

Knight, D. (1914). *An outline of psychobiology*. John Hopkins Press.

Marx, K. (1977). *A Contribution to the Critique of Political Economy*. Moscow: Progress Publishers.

Marx, K. (1977). A Contribution to the Critique of Political Economy. Progress Publishers.

Pinel, J. P. J. (2003). Biopsychology (5th ed.). Allyn and Bacon.

Pinel, J. (2009). *Psychobiology* (7th ed.). Pearson/Allyn and Bacon.

Robert, M. Y. (1996). The mind–body problem. In Companion to the History of Modern Science. Taylor and Francis.

Robinson, H. (2011). Dualism. The Stanford Encyclopedia of Philosophy.

Russell, S., & Norvig, P. (2010). *Artificial Intelligence: A Modern Approach* (3rd ed.). Prentice Hall.

Skirry, J. (2016). *Rene Descartes: The Mind-Body Distinction*. Internet Encyclopedia of Philosophy.

Squire, L. R. (2008). *Fundamental neuroscience* (3rd ed.). Academic Press.

Teufel, C., Ghazanfar, A. A., & Fischer, J. (2010). On the relationship between lateralized brain function and orienting asymmetries. *Behavioral Neuroscience*, *124*(4), 437–445. doi:10.1037/a0019925 PMID:20695643

KEY TERMS AND DEFINITIONS

Behavioral Neuroscience: Also known as biological psychology, biopsychology, or psychobiology, is the application of the principles of biology to the study of physiological, genetic, and developmental mechanisms of behavior in humans and other animals.

Biological Psychology: Is concerned primarily with the relationship between *psychological* processes and the underlying physiological events or, in other words, the mind-body phenomenon.

Biological Techniques: Biological techniques are methods or procedures that are used to study living things. They include experimental and computational methods, approaches, protocols, and tools for biological research.

Biopsychology: Is a branch of psychology that analyzes how the brain, neurotransmitters, and other aspects of our biology influence our behaviors, thoughts, and feelings

Mind-Body Relationship: Typically, humans are characterized as having both a mind (nonphysical) and body/brain (physical). This is known as dualism. Dualism is the view that the mind and body both exist as separate entities. Descartes/Cartesian dualism argues that there is a two-way interaction between mental and physical substances

Physiological Psychology: *Physiological psychology* is a subdivision of behavioral neuroscience (biological *psychology*) that studies the neural mechanisms of perception and behavior through direct manipulation of the brains of nonhuman animal subjects in controlled experiments.

Chapter 2
Anatomy of the Brain

Sumit Kumar
Jawaharlal Nehru University, India

ABSTRACT

Neuroanatomy is a specific branch within neuroscience that deals with brain anatomy. Its broad area includes the brain structure, organization, and localization/networks of the nervous system. It also helps in understanding the sensorimotor systems, along with associated function like learning, behavior, vision, attention, language, and so on. In the present chapter, the author comprehensively discussed the brain basic morphology, architecture, and also some functional aspects of the brain. At the end of this chapter, the author included the tool and techniques used in the study of brain anatomy so that student can learn and understand the topic clearly.

CHAPTER OUTLINE

- Introduction
- Approaches used to study the brain region
- General organization of the brain
- Cerebrospinal fluid
- Choroid plexuses
- Blood-Brain Barrier
- Ventricles of the brain
- Brain anatomy at tissue level
- Brain anatomy at organ level
- Orientation in neuroanatomy
- Tool
- Conclusion

DOI: 10.4018/978-1-7998-2860-0.ch002

Learning Objectives

- Describing the approaches and tool to study brain.
- To describe the location of different brain region.
- Discussion about the Blood-Brain Barrier and blood supply system into brain.

INTRODUCTION

Brain is bilaterally symmetrical, soft, gelatinous structure weighting around 400 grams in infants and 1250-1450 grams in adults. The females have lesser brain weight than male. However, it does not mean that females are lesser intelligent than male since the relation between brain weight and intelligent is already disproved multiple studies. Brain is surrounded by meninges and then enclosed bya bony structure known as cranium. Brain is formed from five region namely telencephalon, diencephalon, mesencephalon, metencephalon, and myelencephalonfrom rostral to caudal region during embryogenesis. However these regions disappearwith development and new structure emerges in the adult brain.

APPROACHES TO THE STUDY OF BRAIN REGION

Brain is anatomically divided into multiple areas based on associated function. Hence it is essential to characterize the brain anatomically. Microscopic examination of animal specimens has been the basic foundation in neuroanatomy. Franz Josef Gall, a German neuroanatomist in the early 1800s, was the first person who tried to locate the brain function (Figure 1a). The studies of Gall gave birth of the new discipline called"phrenology". Gall led down the basic tenets of phrenology: "1) brain is the organ of the mind; 2) The mind is composed of multiple distinct, innate faculties; 3) Because they are distinct, each faculty must have a separate seat or "organ" in the brain; 4). The size of an organ, other things being equal, is a measure of its power; 5). The shape of the brain is determined by the development of the various organs; 6). As the skull takes its shape from the brain, the surface of the skull can be read as an accurate index of psychological aptitudes and tendencies" (Figure 1b, britishlibrary.net). Soon, phrenology transformed itself from "anatomy" to "skull study". Paul Broca in 1860s, successfully identified the brain region for language (Clower et al., 2001). Then experimentation on animals and humans (patients with focal brain lesions, behavioral studies, post-mortem anatomy, structural imaging, and functional imaging) are used to functionally characterize and ascribe the specific function to the specific brain area. Later, physiological response, tract-tracing,

Figure 1. a. Franz Josef Gall (1758-1828) portrait, b. Anphrenology chart

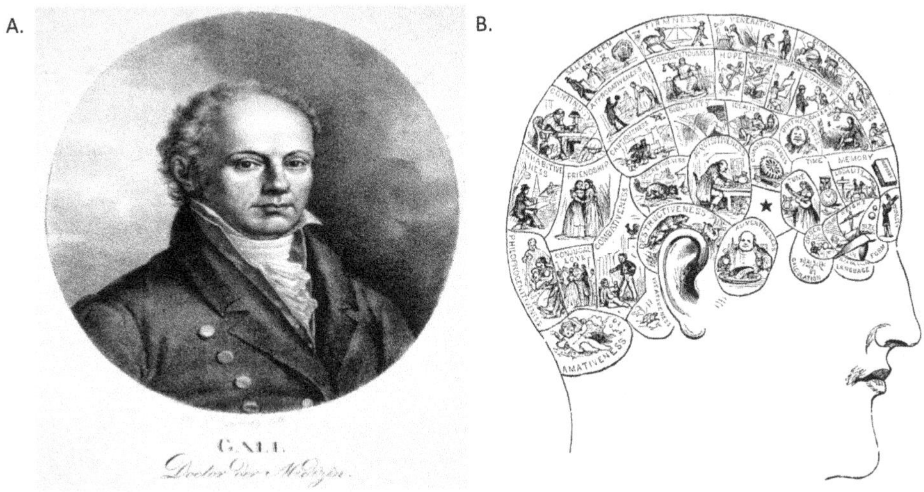

single-unit recordings, behavioral studies, pharmacology, etc. gave the foundation to modern neuroanatomy.

GENERAL ORGANIZATION OF THE BRAIN

Brain is composed of wrinkled, pinkish-gray tissue (Figure2a). It is protected by the thick bony skull; then three thin membranes called meninges and cerebrospinal fluid (Figure2a). Meninges protect the brain, blood vessels, and form an enclosure around venous sinuses. Three connective tissue membranes lie external to the brain aredura mater, arachnoid mater, and pia mater (Figure2a).

Dura Mater: Dura mater is a leathery, strong meninx composed of two fibrous connective tissue layers. The two layers separate in certain areas called dural sinuses.

Arachnoid Mater: It is present at middle meninx formed by loose brain covering. The subdural space separatesarachnoid mater from dura mater layer. The subarachnoid space is present between arachnoid and piamateris filled with cerebrospinal fluid and large blood vessels. Arachnoid villi protrude superiorly and permit cerebrospinal fluid to be absorbed into venous blood.

Pia Mater: Deep meninx composed of delicate connective tissue and clings tightly to the brain.

Figure 2. a. The cerebrospinal fluid and meninges, namely dura mater, arachnoid mater, and pia mater, b. blood-brain barrier.

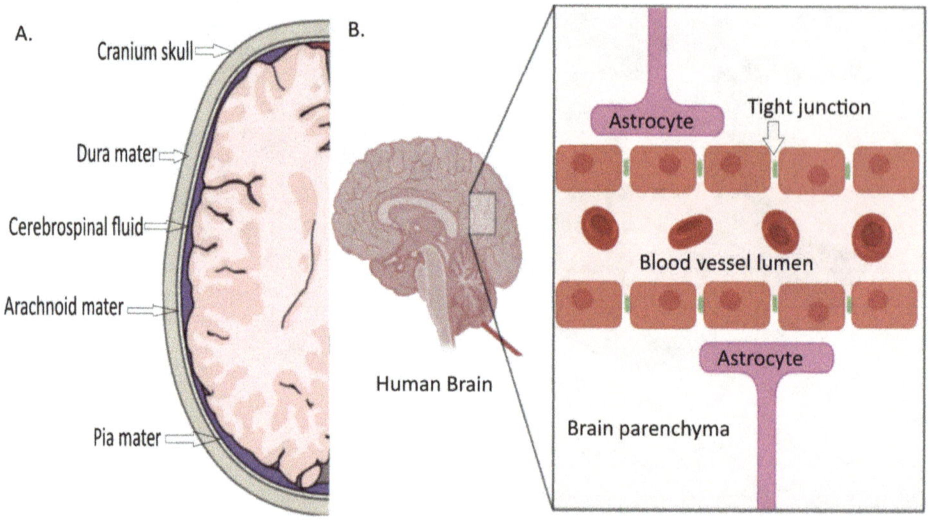

Cerebrospinal Fluid: Cerebrospinal fluid is80-150 ml watery solution similar to blood plasma in composition with lesser protein and different ions concentration. Cerebrospinal fluid forms a liquid cushion that gives buoyancy to the central nerves system organ and thus avoids the brain to crush under its own weight (Figure2a). It protects the central nerves system from blows and other trauma. It also provides nourishment to the brain and carries chemical signals throughout it.

Choroid Plexuses: These are clusters of capillaries that form tissue fluid filters, which hang from the roof of each ventricle. It has numerous ion pumps that allow it to alter ion concentrations of the central nerves system and help in cleaning the central nerves system by removing wastes.

Blood-Brain Barrier: Blood-Brain Barrier is a protective mechanism that helps inmaintenance of stable environment for the brain. The harmful substances are shielded from the brain by the blood-brain barrier. It avoids exposer of blood-borne substances and other toxic chemicals to brain by differential permeability (Figure 2b). Human brain utilizes 25 percent of the total glucose and about 75 percent of oxygen; therefore, blood is pretty important to the normal functioning of brain. The reduction in blood supply causes brain stroke. The selective barrier allows nutrients to pass freely and so ineffective against substances that can diffuse through plasma membranes. Stress increases the permeability of the membrane, and thus even harmful chemicals pass through the blood-brain barrier.

Figure 3. The anatomical positioning of the ventricles in brain.

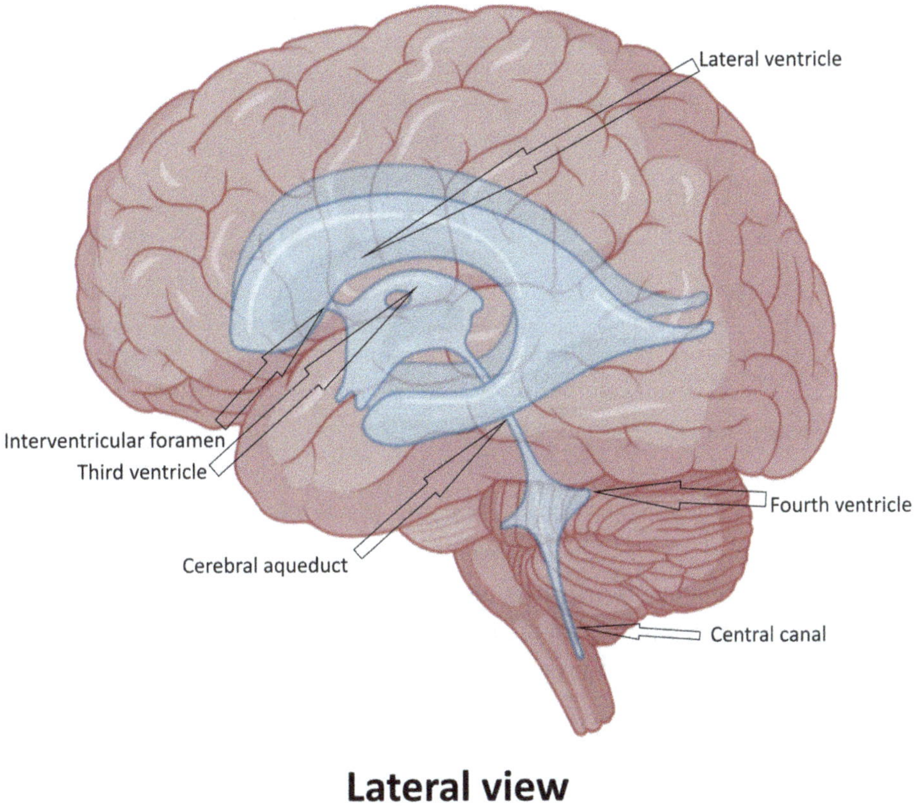

VENTRICLES OF THE BRAIN

The ventricles of the brain are an interactive web of cavities filled with cerebrospinal fluid. Ventricular system is also involved in the production of cerebrospinal fluid and located within the brain parenchyma. It consists of two lateral ventricles (one in each hemisphere), third ventricle (in slit between thalamic halves) and fourth ventricle in between brain stem and cerebellum (Figure 3). The fourth ventricle consist foramen of Monro (the channel between lateral and third ventricles), cerebral aqueduct (the channel between 3rd and 4th ventricle) and central & lateral apertures (4th ventricleand other spaces).

Brain Anatomy at Tissue Level: Nerves system at tissue level is made up of a large mass of nerve tissue, consisting of neurons, glial cells, and extracellular matrix. There are multiple types of neurons, and glial, fulfilling different function. Neurons are the main information-processing and transmitting/communicating

Figure 4. a. Brain division through interhemispheric fissure, b. Human brain lobes.

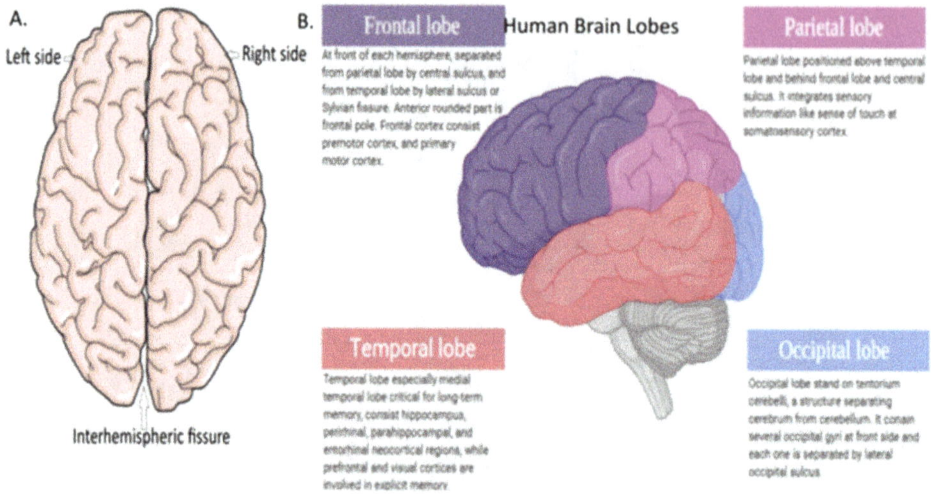

unit in the brain by electrical signals and chemicals called neurotransmitters. Glial cells are responsible for maintenance of homeostasis, myelin production, and provide support and protection for the brain's neuronal cells.

Brain Anatomy at Organ Level: At the organ level, brain is composed of multiple units or regions. These regions are often modular and serve a specific role within the defined pathways of the nervous system. Brain is divided into two halves, namely left and right through an interhemispheric fissure (Figure4a). The central sulcus runs down & forward, while the lateral fissure runs backward & up. This division segregates the brain into four regions, namely frontal, temporal, parietal, and occipital lobes (Figure4b).

Based on function, brain is divided into multiple regions as follow:

1. Brain Stem
    a. Medulla oblongata
    b. Pons
    c. Midbrain (mesencephalon)
2. Diencephalon
    a. Thalamus
    b. Hypothalamus

Figure 5. a. The three distinctive parts of brainstem, marked in different color, b. Human Limbic System.

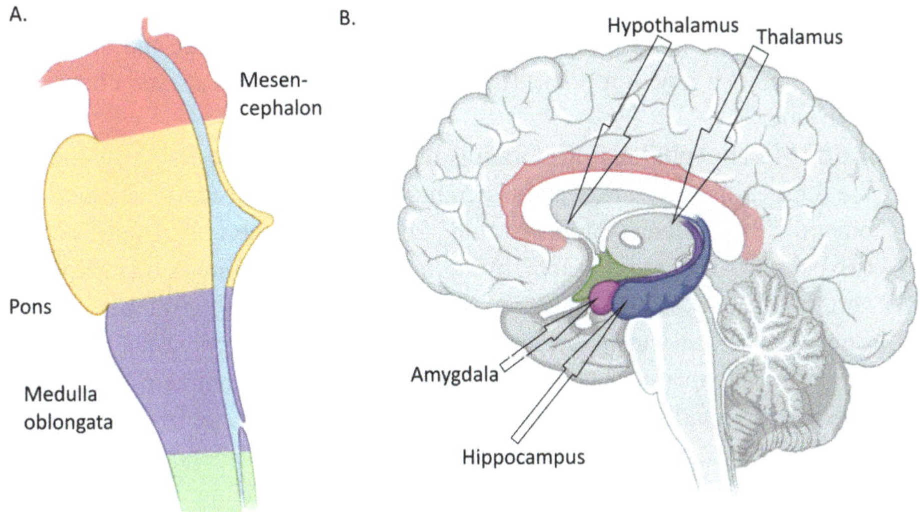

3. Cerebrum
4. Cerebellum

Brain Stem: Brainstem is the oldest or earliest part of the brain. It begins at spinal cord swells and enters the skull. It is responsible for automatic survival functions (Figure 5a). The reticular formation is a netlike network of neurons. The reticular formation is present across brain stem is to filters incoming stimuli and passes the required information to other higher areas of the brain. It is also integrated with 10 of 12 pairs of cranial nerves.

Medulla Oblongata: The medulla is the lowest part of the brainstem (Figure 5a). It has two major tracts of nerve fibers: pyramids and decussation of the pyramids (Iordanova et al., 2019). The pyramids have two longitudinal ridges formed by corticospinal tracts, while decussation of the pyramids is at crossover points to other side of the spinal cord for the corticospinal tracts.

It also contains inferior olivary nuclei, vestibular nuclear complex, ascending sensory tract nuclei, cardiovascular control center, and respiratory centers. The inferior olivary nuclei transfer signals from the spinal cord to the cerebellum and regulate the learning process. The vestibular nuclear complex is responsible for controlling eye movement. The force and rate of heart contraction is regulated by the cardiovascular control center. Similarly, the frequency and intensity of breathing is regulated by the respiratory centers present in medulla.

Pons: It is involved in coordinating movement, sleeping, waking, and dreaming. Brainstem is a junction point, from where most nerves pass and connect to each side.

Midbrain (Mesencephalon): Midbrain is present between the diencephalon and the pons (Figure 5a). It contains cerebral peduncles, cerebral aqueduct, and other different nuclei (Caminero et al., 2019). The cerebral peduncles are the two bulging structures that contain descending pyramidal motor tracts from the cortex. The third and fourth ventricles connect through the hollow tube known as cerebral aqueduct. It also consisted of four nuclei, namely corpora quadrigemina, superior colliculi, inferior colliculi, and substantianigra(Caminero et al., 2019). The corpora quadrigemina is four dome-like protrusions coming out of dorsal midbrain. The superior colliculi form visual reflex centers, while inferior colliculi form the auditory relay centers. The substantianigra is functionally linked to basal nuclei. It is the same area that affected and responsible for Parkinson's disease.

Diencephalon: It is the central core of the forebrain that encloses the third ventricle. It consists of three paired structures, namely thalamus, hypothalamus and epithalamus (with pineal gland).

The Limbic System: Limbic System is a doughnut-shaped system of neural structures at the border of the brainstem and cerebrum (Figure 5b). It includes the hippocampus, amygdala, and hypothalamus (Figure 5b). It is associated with emotions such as fear, aggression, and drives for food and sex.

Thalamus: Thalamus is paired, egg-shaped masses that form the superolateral walls of the third ventricle (Figure 5b). It contains four groups of nuclei, namely anterior, ventral, dorsal, and posterior. Its primary function is to converge and synapse the all afferent impulses in thalamus. All inputs going to the cerebral cortex are cross through the thalamus. Its main function is sensation, motor activities, cortical arousal, learning, and memory.

Hypothalamus: It is sitting atop of brainstem (Figure 5b). It receives information from all senses (except smell) and routes it to the higher brain regions that deal with seeing, hearing, tasting, touching. It acts as a junction point from where all communication going pass to various destinations in the higher brains region. It is a switchboard, which passes the messages to the sensory part in the cortex, cerebellum, and medulla. It controls and regulates many functions such as hunger, emotions, body temperature, heartbeat, sexual behaviors, digestive tract mobility, eating, drinking, and thirst. It also regulates the endocrine function through the pituitary gland.Hypothalamus also acts as a reward center which is located at "pleasure centers". In the 1950s, two psychologists named James Olds and Peter Milner implanted the stimulating electrode in rats at the pleasure center. Rats were found to press the paddle as many as 7000 times per hours for stimulating pleasure center rather than paddle for food and water (Olds et al., 1954). This was happening even when rats were hungry or thirsty. Humans have limbic area for pleasure. Stimulation of limbic area in human has shown a mild pleasure; unlike rat experimentation where rats became frenzy (Zald et al., 1998).

Amygdala: Amygdala is two lima bean-sized neural clusters (Figure 5b). It influences aggression, fear process and also responsible for the processing of emotional memories.

Cerebral Hemisphere: The cerebral hemisphere is the topmost or superior part of the brain and forms 83% of total brain volume (Figure 6). It consists of ridges, groove /valley and fissure termed asmass gyros (gyri),sulcus (sulci) and long deeper groovesrespectively (Figure 6). Cerebral hemispheres are the most advance structure and strongly associated with logic and complicated task (Bui et al., 2019). Thus it is the highest evolved in human species. The gyri and sulcus increase the surface area of the brain and therefore provide an expansion to the brain for advancement without increasing the volume.

Cortical Region: It is 0.125 inches thick and consists of 40% of brain mass. The cortex is the superficial gray matter (Figure 6). It is consists of motor areas, sensory areas, association areas, and contralateral control, enabling sensation, communication, memory, understanding, and voluntary movements. However, no functional area act alone, and thus conscious behavior is an outcome of involvement of the entire cortex.

The cortex is divided into two hemispheres, and each hemisphere is connected via the corpus callosum. Corpus callosum is a thick bundle of nerves fiber, transmitting the information from one side to other. The corpus callosum is cut in schizophrenic patients to prevent or reduce the seizure. It is not present in all organisms; in fact it is just a small bundle of connections in birds that allow their two hemispheres to communicate. Each hemisphere acts contralaterally means one side controls the opposite side of the body. Each hemisphere is not equal in function, which is depends on left or right-hand uses.

The cortical region contains three types of functional areas, namely: motor region, which controls voluntary movement, sensory region controlling consciousness and wakefulness, and association region, which assimilate diverse information.

Figure 6. Human brain cortical region

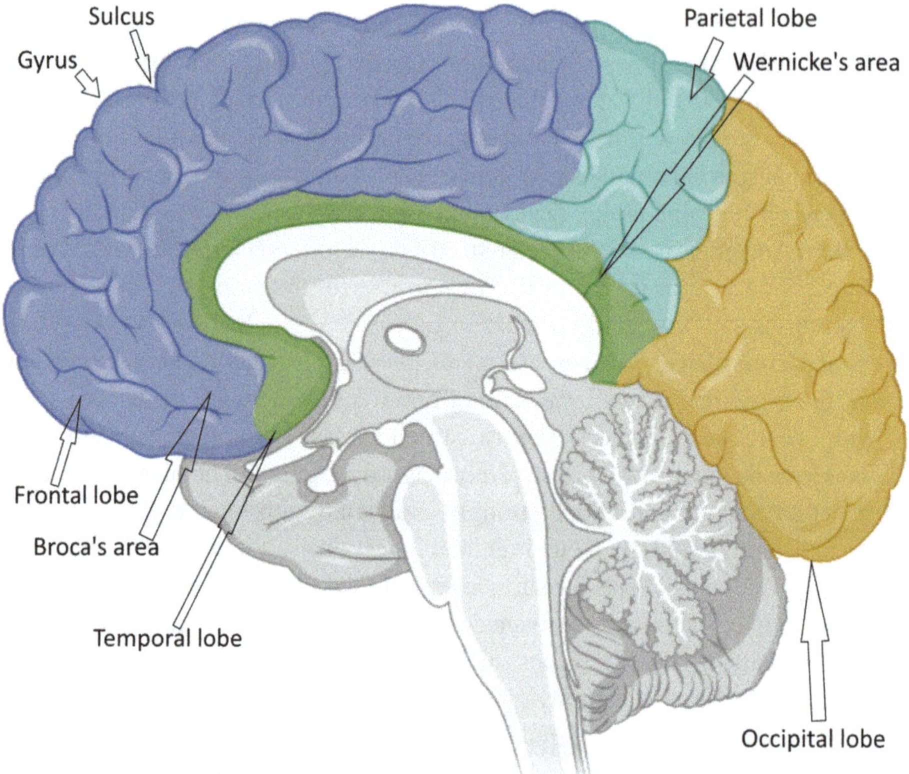

Motorregion: Motor region is the primary motor cortex consisting of pyramidal cells and forming corticospinal tracts (axonal tract). It is located in the precentralgyrus. It is responsible for the conscious control of preciseand skilled voluntary movements (Barthas et al., 2017).

Premotor Cortex: It controls and responsible for the learned, repetitious, or patterned motor skills task. It is also participating in coordinates simultaneous movement or sequential actions(Barthas et al., 2017).

Frontal Eye Field: It is present at the anterior location in the motor region. It controls the voluntary eye movement(Barthas et al., 2017).

Broca's Area: The Broca's area present near lateral sulcus on the frontal lobe (Friederici., 2018). It is present in usually left hemisphere (sometimes in the right), governing motor speech region which instructs tongue muscles to active as one prepares to speak(Friederici., 2018).

Sensory Areas: It divided into five regions, namely: 1). Primary somatosensory cortex; 2). Somatosensory association cortex; 3). Wernicke's area; 4). Visual areas; 5). Auditory areas.

Primary Somatosensory Cortex: It is located in the postcentralgyrus. It obtains signals from the skin and skeletal muscles.

Somatosensory Association Cortex: It integrates sensory information and forms a comprehensive understanding of the stimulus like size, texture, and relationship of parts.

Wernicke's Area: It has the ability to sound out new words and sounds (Binder., 2017).

Visual Areas: It is primary visual cortex, located at the extreme posterior tip of the occipital lobe (Tamietto et al., 2018). It receives visual information from the retinas and interprets the visual stimuli such as color, form, and movement.

Auditory Areas: It is primary auditory cortex, located at the superior margin of the temporal lobe. It obtains information related to pitch, rhythm, and loudness.

Association Area: It consists of prefrontal cortex and language areas (Tomasi et al., 2019). The prefrontal cortex is located in the anterior portion of the frontal lobe and involved with intellect, cognition, recall, and personality. The prefrontal cortex is essential for judgment, reasoning, persistence, and conscience. It tightly associated with the limbic system. The language area is located in a large area surrounding the left lateral sulcus. It consists of Wernicke's area and Broca's area, which we already discussed above.

Cerebellum: Cerebellum is attached to the posterior side of the brainstem (Figure 7). It is also known as "little brain" (Bray., 2017). It helps coordinate voluntary movements and balance. It is also involved in nonverbal learning, and memory.

ORIENTATION IN NEUROANATOMY

In neuroanatomy, many specific terms are used to define the orientation and location along with brain axis. Dorsal and ventral region is the most commonly used term and refers to the top or upper side and bottom or lower side, respectively (Hendelman et al., 2005). Rostral and caudal refer to the front of the body and tail end of the body, respectively. Medial and lateral refer to the region, close, or relatively closer, to the midline and a location more or less separated away from the midline.

Tools: Advancement in neuroanatomy field is directly related to the technologies used to perform investigations. Hence, it is obligatory to know the different tools which are used in neuroanatomy research.

Figure 7. Anatomy of the cerebellum

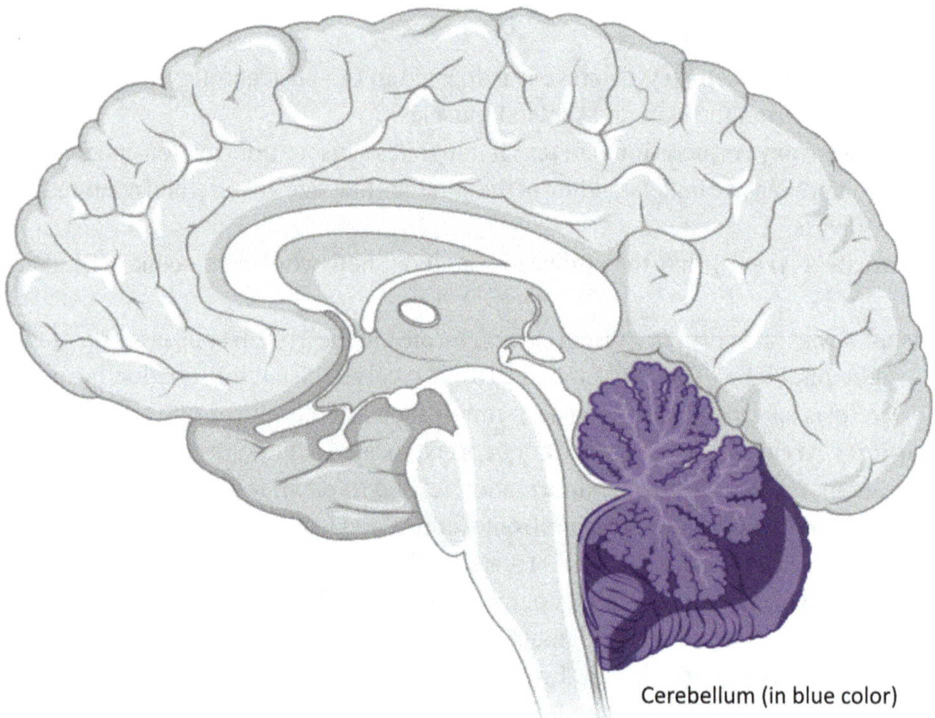

Cell Staining: Staining is essential to understand the neuronal placement and easily visualized the brain features at the microscopic level. Nissl staining has been used to stain the plentiful acidic polyribosomes in the rough endoplasmic reticulum using basic aniline dyes(Gittins et al., 2004). The stained portion is called as Nissl bodies. It allows the investigator to differentiate the different cell types like neurons and glia, and their shapes, cytoarchitecture, and also sizes. Golgi's method is based on silver staining technique, originally described by Camillo Golgi in 1873 (Golgi,1973). It uses potassium dichromate and silver nitrate and fills the neurons, glia with black-brown silver chromate precipitate. It is widely used to stain neuronal cell bodies, neurites of some neurons –dendrites.

Histochemistry: Histochemistry is another technique, where biochemical reaction is used to visualize the brain microscopically (Lopes., 2019).

Non-Invasive Brain Imaging: Computed tomography, Diffuse optical tomography, Diffusion tensor imaging, Magnetic resonance imaging, Single-photon emission computed tomography, Positron emission tomography, and Functional magnetic resonance are the common techniques that are routinely used with radioactive and non-radioactive tracers to study the brain structure and function non-invasively (Orrison et al., 2017).

The other common methods are dye-based methods for tracing the axonal path,connectomics to visualize serial brain sections under the microscope, computational neuroanatomy to model and quantify the spatiotemporal dynamics of neuroanatomical structures etc. are used to study neuroanatomy.

CONCLUSION

From the present chapter, it's clear that brain anatomical study is the critical and initial point in understanding the brain function, disorder, and therapeutic opportunity. Neuroanatomy represents a tremendous opportunity to unlock the mysteries of the brain and its functions. With the emergence of advanced technologies and computational informatics, allowing investigators to realize the full potential of the electronic resources, tool, and methodologies for brain exploration and thus ensuring the society for receiving the maximum possible benefits out of neuroscience research. The human Brain Mapping Initiative is one of the first steps in such direction.

REFERENCES

Barthas, F., & Kwan, A. C. (2017). Secondary motor cortex: Where 'sensory' meets 'motor' in the rodent frontal cortex. *Trends in Neurosciences*, *40*(3), 181–193. doi:10.1016/j.tins.2016.11.006 PMID:28012708

Binder, J. R. (2017). Current controversies on Wernicke's area and its role in language. *Current Neurology and Neuroscience Reports*, *17*(8), 58. doi:10.100711910-017-0764-8 PMID:28656532

Bray, N. (2017). Cerebellum: The little learning brain. *Nature Reviews. Neuroscience*, *18*(5), 263. doi:10.1038/nrn.2017.47 PMID:28381834

Bui, T., & Nguyen, J. (2019). Neuroanatomy, Cerebral Hemisphere. StatPearls Publishing.

Caminero, F., & Cascella, M. (2019). Neuroanatomy, Mesencephalon Midbrain. StatPearls Publishing.

Clower, W.T., & Finger, S. (2001). Discovering trepanation: the contribution of Paul Broca. *Neurosurgery, 49*(6), 1417-26.

Friederici, A. D. (2018). The neural basis for human syntax: Broca's area and beyond. *Current Opinion in Behavioral Sciences*, *21*, 88–92. doi:10.1016/j.cobeha.2018.03.004

Gittins, R., & Harrison, P.J. (2004). Neuronal density, size and shape in the human anterior cingulate cortex: a comparison of Nissl and NeuN staining. *Brain Research Bulletin, 63*(2), 155-60.

Golgi, C. (1873). Sulla sostanzagrigiadelcervello. *GazettaMedicalItaliana.*, *33*, 244–246.

Hendelman, W. (2005). *Atlas of functional neuroanatomy*. CRC Press. Retrieved from http://pages.britishlibrary.net/phrenology/

Iordanova, R., & Reddivari, A. K. R. (2019). Neuroanatomy, Medulla Oblongata. StatPearls Publishing.

Lopes, M. B. S. (2019). Diagnostic histochemistry in neuropathology. *Seminars in Diagnostic Pathology*, *36*(1), 71–82. doi:10.1053/j.semdp.2018.10.004 PMID:30471893

Olds, J., & Milner, P. (1954). Positive reinforcement produced by electrical stimulation of septal area and other regions of rat brain. *Journal of Comparative and Physiological Psychology*, *47*(6), 419–427. doi:10.1037/h0058775 PMID:13233369

Orrison, W.W., Lewine, J., Sanders, J., & Hartshorne, M.F. (2017). Functional brain imaging. *Elsevier Health Sciences, 24*.

Tamietto, M., & Leopold, D. A. (2018). Visual cortex: The eccentric area prostriata in the human brain. *Current Biology*, *28*(1), R17–R19. doi:10.1016/j.cub.2017.11.006 PMID:29316413

Tomasi, D., & Volkow, N. D. (2019). Association between brain activation and functional connectivity. *Cerebral Cortex*, *29*(5), 1984–1996. doi:10.1093/cercor/bhy077 PMID:29668865

Zald, D.H., Lee, J.T., Fluegel, K.W., &Pardo, J.V. (1998).Aversive gustatory stimulation activates limbic circuits in humans. *Brain: A Journal of Neurology, 121*(6), 1143-54.

KEY TERMS AND DEFINITIONS

Blood Brain Barrier: Blood brain barrier is a protective element with semipermeable architecture formed by extracellular fluid that separates the circulating blood from the brain and extracellular fluid in the central nervous system. Blood brain barrier system allows passage of small molecules including gases such as oxygen and carbon dioxide by passive diffusion; as well also promote the selective transportation of essential nutrients or molecules such as, water, glucose and amino acids, which cannot diffuse due to absence of required polarity.

Brain: The brain is one of the largest, most vital and complex organs present in human body. It is made from more than 100 billion nerves that forms trillions of connections or projection subsequently called synapses. Synapses pass message from one location to another location via chemical transmission. It is located in the head, close to the sensory organs for senses such as vision.

Brain Anatomy: The brain is divided into 3 parts namely cerebrum, brainstem, and cerebellum. Brain stem is placed at lowest position extending from upper cervical spinal cord to diencephalon of the cerebrum. Cerebrum is the largest part of the human brain, and associated with higher order of function such as thought and action. Cerebellum, also known as "little brain", is divided into two highly folded hemispheres namely anterior and posterior lobe.

Cerebrospinal Fluid: Cerebrospinal fluid is a clear, colorless body fluid surrounding the hollow space present in brain and spinal cord. It is produced by ependymal cells, a specialized group of cells present in the choroid plexuses of the ventricles. It provides cushion to the brain from shocks, supports the venous sinuses, and also play an important role in the brain homeostasis and metabolism.

Cerebrum: Cerebrum or telencephalon is the largest part of the brain. It is divided into approximately two symmetric halves, known as cerebral hemispheres. It controls emotions, thought, speech, hearing, vision, personality, reading, learning, etc.

Cortex: Cortex is a wrinkly structure of tightly packed neuron surrounding the brain. It is responsible for higher thought processes including speech, decision making process, writing, composing sentences, and problem-solving. It is divided into four different lobes namely frontal, parietal, temporal, and occipital lobe. Each lobe is responsible for processing different types of sensory information.

Neuroanatomy: Neuroanatomy is the study of brain organization and other functionally segregated structure present in nervous system. In vertebrates, it is divided into central nervous system and peripheral nervous system consisting brain and spinal cord and the nerves projection that connect it to the rest of the body.

Ventricle: It is a cerebrospinal fluid filled cavity present at brain parenchyma. The fluid is produced by choroid plexuses, which is located in the ventricles. It consists, two lateral ventricles, a third ventricle, cerebral aqueduct, and the fourth ventricle.

Chapter 3
Structure and Function of Cells

Shazia Ali
Sher-I-Kashmir Institute of Medical Sciences (SKIMS), India

Sumit Kumar
Jawaharlal Nehru University, India

Asma Ali
Independent Researcher, India

Syed Irtiza
Sher-I-Kashmir Institute of Medical Sciences (SKIMS), India

ABSTRACT

Neurons are the elementary structures that process information in the CNS. They are electrically excitable cells that process and transmit information around the nervous system. Neurons communicate information either by electrical or by chemical signaling that occurs through synapses. Neurons are the main part of nervous system consisting of a number of specialised types. Neuronal structure is similar to other body cells, having common features: Neurons are enclosed in a cell membrane known as a plasma membrane. The nucleus of a neuron compromises of chromosomes and genetic information. Neurons comprise of cytoplasm, mitochondria, and other organelles. The cellular processes take place in a neuron. Ribosomes are used for production of proteins, and mitochondria are for metabolic action acting as powerhouse of cell. Neurons comprise a Golgi complex that has a system of vesicles that secretes hormones and other products. The difference in neurons or nerve cells from other body cells is that extensions originate from the central body of the neuron.

DOI: 10.4018/978-1-7998-2860-0.ch003

NEURONS

Neurons are the elementary structures which process information in the Central Nerves System (CNS). They are electrically excitable cells that can receive and transmit chemical or electrical signals rapidly through synapses around the nervous system. Neurons are the main part of nervous system consisting of number of specialised type and function. Neuronal structure is similar to other body cells, having similar features, yet there is complexities vary throughout animal kingdom. It change from nerve nets (neurons) in lower organism to centralized structure consisting central nervous system (CNS) made of brain, spinal cord and a peripheral nervous system (PNS) made up of peripheral sensory and motor nerves. One major difference between invertebrates and vertebrates nervous systems is that nerve cords in various invertebrates are ventrally placed where dorsally vertebrate. They are enclosed in a cell membrane known as a plasma membrane. The nucleus of a neuron compromises of chromosomes that possess genetic information. Neurons comprises of cytoplasm, mitochondria and other organelles. Basic cellular processes take place in a neuron. Ribosomes are involved in protein production, while mitochondria act as power house of cell by metabolic action. Neurons comprise a Golgi complex which has a system of vesicles that secretes hormones and other products the difference in neurons or nerve cells from other body cells is the extensions originate from the central body of the neuron. These are termed as dendrites and axons. Dendrites pass information to the cell body while axons communicate information away from the cell body. An electrochemical mechanism occurs transmitting and receiving information from neurons There are distinct and specialised structures, synapse (the junction between one neuron and the next) which are compromised by neurons. Neurotransmitters and neuro modulators are chemicals synthesized by neurons.

NEURON STRUCTURE

A distinctive neuron possesses a cell body (soma), dendrites and axon. Dendrites are filaments in nature, originate from the cell body, they branch several times forming a complex dendritic tree. Small protrusions present in dendrites called dendritic spines, which further increase the surface area require for synaptic connection. An axon is like rob in an electrical cable. The axon mound is the site where cell body starts and explorations to the site in the nervous system connecting with another nerve cell or different type of cell, such as muscle (Fig. 1).

The cell body possess cytoplasm as like other cells with granules called as Nissl's granules. It frequently gives rise to several dendrites but only to one axon formation. A layer of fatty cells called the myelin sheath segmentally encloses the fibres of

Figure 1. Neuronal structure

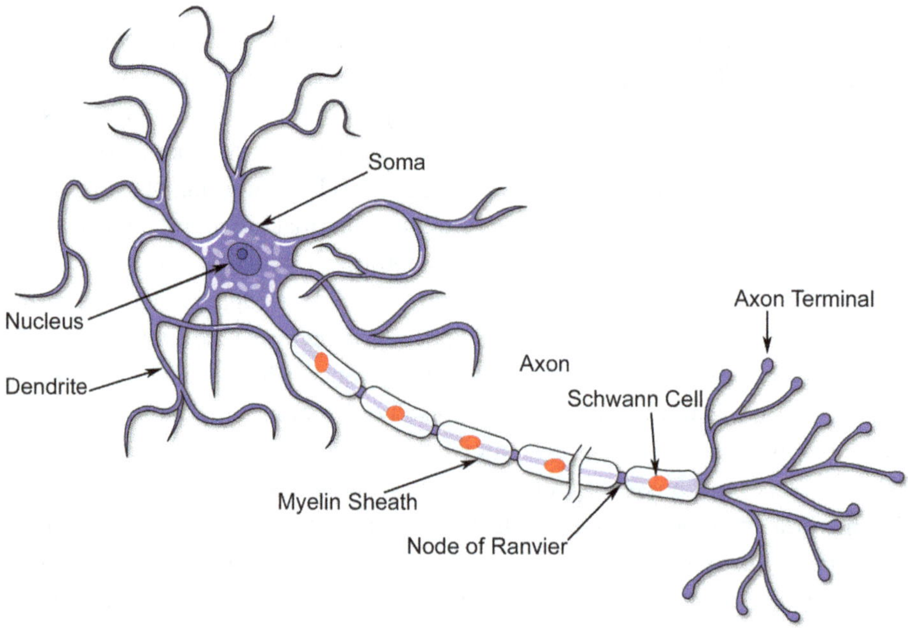

several neurons which increases the transmission speed of neutral impulses. The periodic gaps present along axon over myelin sheath called nodes of Ranvier. It is "recharged" signal as it travels along axon.

NEURON AND ITS CHARACTERISTICS

The neuron compromises of cell body (or soma) that contains a nucleus, smooth and rough endoplasmic reticulum (ER), Golgi apparatus, lysosomes, mitochondria and other cellular components (Fig. 2).

Nucleus contains genetic material packed within the chromosomes comprising information required for all cell function. It transcribes DNA into RNA and then ships it through pores present in nuclear envelope into cytoplasm for protein synthesis. Nucleolus help in ribosomes production and t RNA synthesis required for translate of genetic information into protein. Lysosomes compromises of enzymes which decrease chemicals to their different components. Centrosome is a microtubule network that controls the cells and the cell cycle. Cytoplasm this is a watery and salty fluid with a potassium-rich partially transparent gelatinous solution filling the cell.Vacuoles are sections bound by membranes which are required for storing

Figure 2. Neuron cell ultra structure showing cellular components

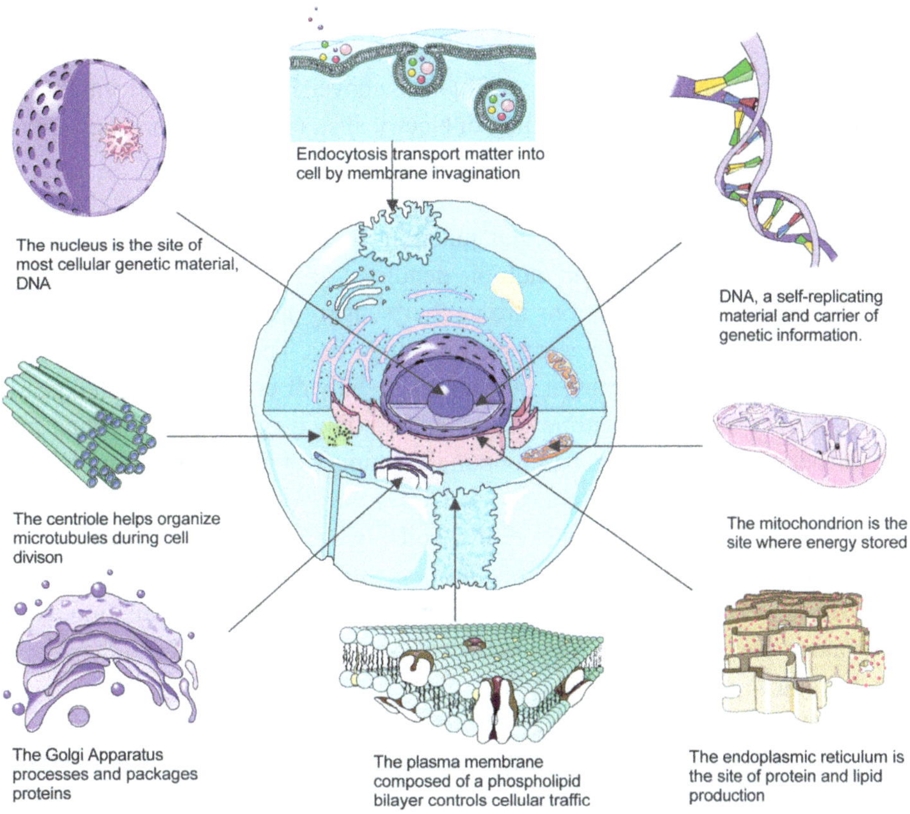

and secretion of neurotransmitters. Endoplasmic reticulum (ER) tubes are used in network to transport materials around the cytoplasm. ER compromising ribosomes are called rough ER, whereas those having no ribosomes are called smooth ER. ER having ribosomes are important for protein synthesis. Golgi body or Golgi apparatus a membrane-bound structure used for processing of peptides, proteins and neurotransmitters into vesicles. Microfilaments or neurotubules helps in transporting materials within a neuron and structural help.

NEURON CLASSIFICATION

In the nervous system, neurons may be categorized according to their function (sensory, motor, interneuron), their locality (cortical, subcortical), the characteristics of the neurotransmitter synthesised and released by them (cholinergic, glutamatergic) and their shape (pyramidal, granule, etc.) On basis of their axons and dendritic cells leaving their cell body or soma, they are categorized into three main type:

1. **Sensory Neurons:** It is responsible for converting stimulus from external environment into internal stimuli or electrical signals. Afterword signal reach to spinal cord or brain.
2. **Motor Neurons:** They are present in CNS and directly or indirectly control muscles through neuromuscular junction. It is multipolar, means each cell contains a single axon and multiple dendrites.
3. **Interneurons:** Interneurons are neither sensory nor motor instead they act like "broker" between sensory and motor neurons. They are multipolar neuron present in CNS and transfer information locally or connect with only nearby neuron. It is just to save time and prevent sending message back to spinal cord.
4. Structural classification based upon the number of extension from cell body (Fig. 3):
5. **Bipolar:** They are related to retinal cells, two progressions spread from the body of bipolar neurons.
6. **Unipolar:** For each unipolar cell, there are two dorsal root ganglion axons. One of the axon expanses out in the course of the spinal cord and the other in the course of the skin or muscles.
7. **Multipolar:** They comprise many processes that branch out from the cell body. There are though neurons having only have one axon (e.g. spinal motor neurons).

Classification Based on Mode of Transmitting Information

1. **Efferent Neurons (Motor Neurons):** The information is send away from the brain towards muscles and glands.
2. **Afferent Neurons (Sensory Neurons):** The information is send to the central nervous system from sensory receptors.
3. **Interneurons:** Send information between motor neurons and sensory neurons in central nervous system.

NEUROGLIAL CELLS

They are simply glial cells that holds nervous system together. Neuroglial cells provide physical integrity and nutrition to the nervous system and preserve homoeostasis. They out number neuron with a ratio of 3 to 1. Some of the glial cells physically support other kind of cells; others regulate the internal environment surrounding

Figure 3. Unipolar, bipolar, and multipolar neurons.

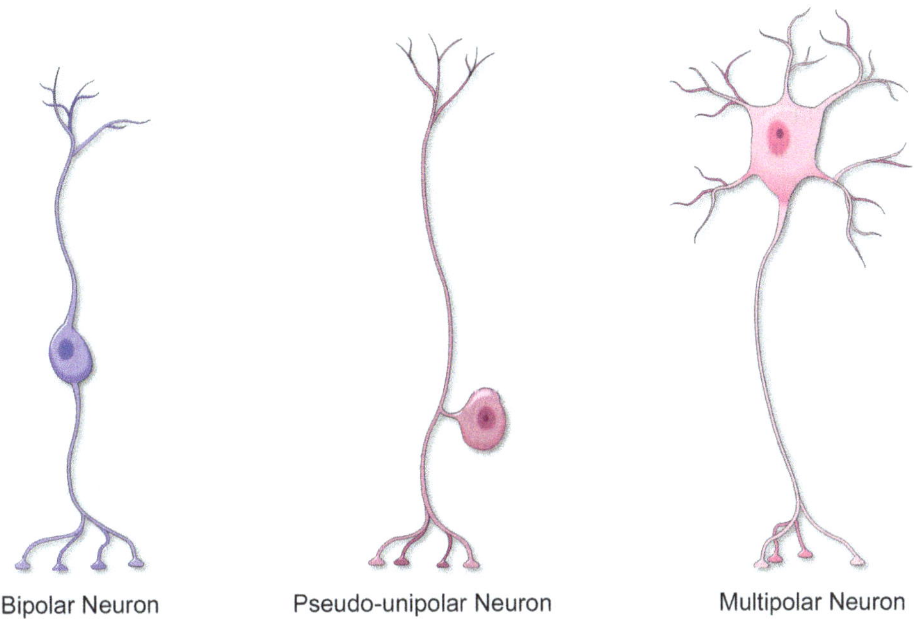

neurons or nerve cells and provide them nutrition. Glial cells are also involved in cortical development, neuronal function, axons and dendrites development, thus play a major role in the growth of the nervous system. Earlier glial cells were thought to be required for only neuronal growth but recently, a new type of glial cell, the oligodendrocyte precursor cell, is involved in synaptic signalling in the hippocampus.

Types of Glial Cells

There are three types of glial cells. Some of them are found in central nervous system and others are important for functioning of the peripheral nervous system.

1. **Astrocytes:** Astrocytes are star like cells present in brain and spinal cord. They are accountable for maintaining the outer chemical environment around nerve cells by disposing of surplus ions, particularly potassium and chloride, and neurotransmitters are recycled and reused which were released during synaptic transmission. A small swellings which surround and support the endothelial cells of blood brain barrier is end feet formed by astrocytes This is a protection for brain from any harmful substances present in the blood although the metabolic function occur normally. The brain contains two dissimilar types of astrocytes: protoplasmic astrocytes having grey matter and fibrous astrocytes

having white matter of spinal cord and brain. Astrocytes are small cells and predominant glial cells, their numbers are originated in different areas of the brain.

2. **Oligodendrocytes:** In the CNS, oligodendrocytes are mounted creating a myelin sheath that insulates the axon and electrical signals are generated and propagated more efficiently. 50 axons are coated by one oligodendrocyte. Myelin speeds action potential sent along an axon. Around the axon, the sheaths of myelin have gaps between them as the nodes of Ranvier. It is at these gaps that ions can cross the membrane and an action potential occurs. Thus the action potential moves from one node to another, enabling it to move signal faster along myelinated axons than unmyelinated counterparts. Reasoning and performance are affected by myelin loss within the CNS. In the PNS, cells that elaborate myelin are called Schwann cells.
3. **Microglia:** Microglia are glial cells that serve the CNS immune system and are also specific macrophages. Microglia are important protectors of the central nervous system that guard and supports the neurons of the CNS. They undergo a process called phagocytosis that clears the waste from the nervous system. Microglial cells are derived from embryologically from haematopoietic precursors, producing all types of blood cell. Approximately 15-17% of cells are microglia found in all parts of the brain and spinal cord. Microglia increase in in growth in response to brain injury. Microglia are thought to be a feature in neurodegenerative disorders like multiple sclerosis, motor neurone disease and Alzheimer's disease. These cells create the inflammatory response to brain trauma and for the HIV-1 virus in the central nervous system, they are major targets.
4. **Radial Glial Cells:** Radial glial cells are a provisional cell population found in developing CNS. They function both as precursor cells and as a scaffolding outline to support neuron migration.
5. **Schwann Cells:** The cell in the PNS which function similar to oligodendrocytes in the CNS, the Schwann cell, insulates lone discrete axon. Schwann cells are specific types of glial cell that deliver myelin insulation to axons in the peripheral nervous system. Non-myelinating Schwann cells are important for neuronal survival and maintenance of axons.

THE ACTION POTENTIAL AND NERVE IMPULSE

A neuron is at resting stage, and a resting membrane potential is present during this period. However, when our nervous system transmits information, an action potential or a nerve impulse occurs. An action potential is a short-range event in which the electrical membrane potential of a cell rapidly rises and decreases. Depolarisation is the process where positive ions passage is decreased in the cell which in turn decreases the negative charge and thus decreases the charge across the entire membrane. When depolarisation occurs at –55 mV (the inclining point) a neuron will no longer have an action potential. The threshold level is important for an action potential. Once the threshold value is reached, an action potential will be generated for any neuron.

Action potential or nerve impulses occur on exchange of ions across the neuron. A stimulus, for example at the touch receptors causes opening of Na channels positive charge ions to move inside a negative charged neuron. The neuron is positive and depolarised. The process is followed by opening of potassium channels and K ions moves out of cells which in turn reverses the depolarisation. Once the Na ions enter the cell, the membrane potential rises, and when K ions exoduses, it decreases, giving rises to the spike potential. Sodium channels returns back to normal closed state, which makes the action potential to revert towards –70 mV (a repolarisation). The membrane potential essentially transfers past –70 mV to nearly –80 mV (a hyperpolarisation) due to the K ion channels remaining open just a bit too extended. Ion concentrations will also arrive to the resting membrane, and the cell will arrive to –70 mV. An absolute refractory period occurs, where the neuron is at resistant to any stimulant and cannot make another action potential. The absolute refractory period heads another brief period, known as the relative refractory period, during which the neuron can create action potential but the stimulus must be of larger intensity than normal (Figure 4)

SYNAPTIC TRANSMISSION

It was found in 1990, Ramón y Cajal detected a thin gap which separates one neuron from the next stating that neurons do not merge physically. In 1906, Sir Charles Scott Sherrington termed it as a synaptic delay (0.3 to 0.5 ms) and found that some kind of communication is present at these gaps between nerve cells. This point of communication is termed as a synapse through which neuronal cells and non-neuronal cells in glands and muscles signal to each other. Synapses make it possible for nerve cells in the central nervous system to create interconnected neural circuits or networks. The synapse has several parts, a synaptic terminal, a nerve cell that communicates information (or the presynaptic ending), a nerve cell that gets the information (the postsynaptic ending), and gap between them known as the synaptic cleft (Figure 5).

Figure 4. Action potential

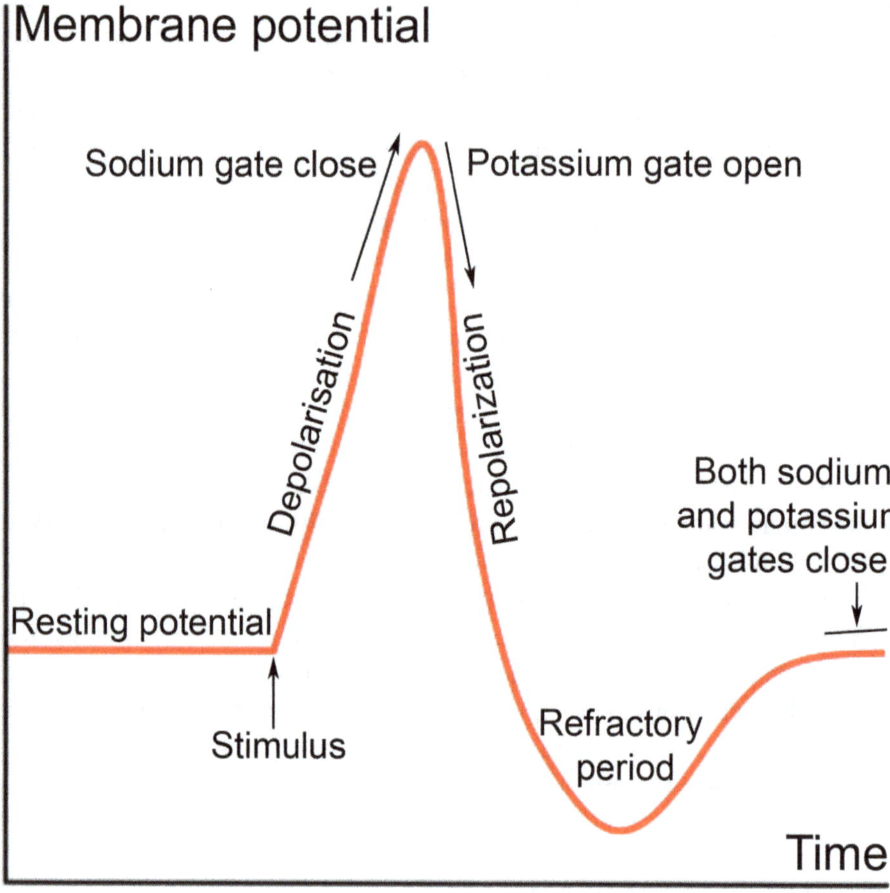

Neurotransmitters: The chemical messenger which are synthesised by neurons across the synapse for communications are called as neurotransmitters. There is a presynaptic ending or synaptic terminal, also known as the synaptic knob which keeps cell organelles like mitochondria and neurotransmitters together. The neuron sending the impulse activates the passage of vesicles which has the neurotransmitter to the membrane of the synaptic terminal. Fusion of vesicle membrane with the presynaptic membrane causes neurotransmitter to release into the synaptic cleft (a gap of 20 nm); where it binds to receptors of another cells causing excitation or inhibition of electrical impulses. The area in the synapse releasing neurotransmitter is called as active zone. Synapse is the area of action and trigger for most of psychoactive drugs. Such neurotoxins change neurotransmitter release and reuptake and access to receptor binding sites.

Figure 5. Synapse

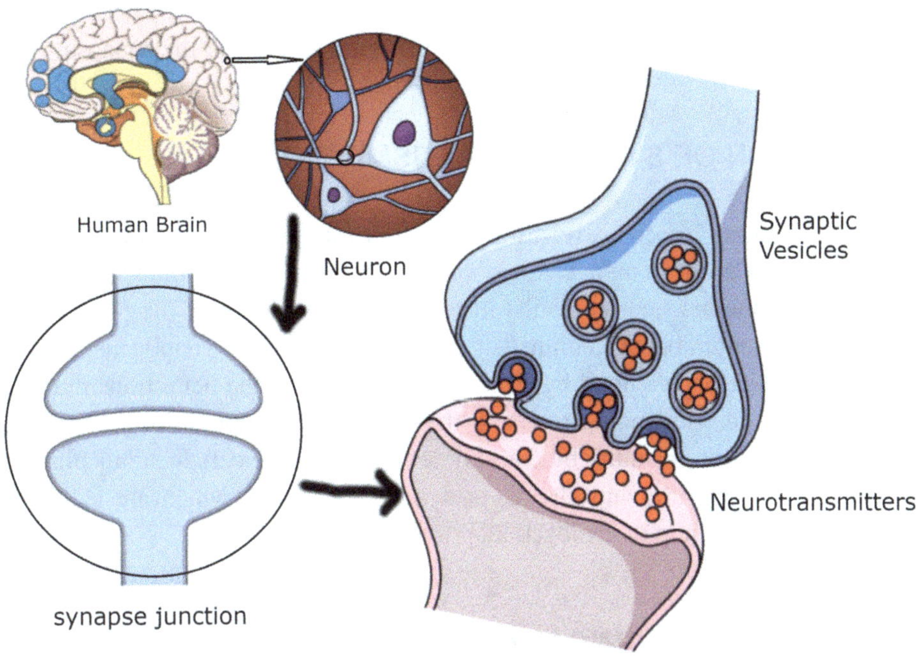

For example, lysergic acid diethylamide (LSD) inhibits with serotoninergic synapses, while cocaine inhibits reuptake of dopamine and its effect is increased. The poison strychnine inhibits the inhibitory effects of the neurotransmitter glycine which causes convulsions and muscle spasm (Bennett & Zukin, 2004).

THE SYNAPTIC VESICLE

Upon activation of presynaptic terminal, calcium channels open and calcium ions enter in the cell. This calcium influx causes the drive and fusion of secretory vesicles to the cell membrane (a process known as exocytosis). In synaptic membrane, vesicles inside the presynaptic cell, neurotransmitters are sealed in. Neurotransmitters are released into synaptic cleft by the inward force of action potential. The neurotransmitter molecules move across the synaptic cleft by dispersal to receptor sites on the presynaptic membrane, opening ion channels in the postsynaptic cell membrane. It leads to movement of ions modifying the cell's local transmembrane potential resulting in a change in the potential of the postsynaptic cell or a postsynaptic potential. It leads to an excitatory reaction as a result of depolarisation or an inhibitory reaction as a result of hyperpolarisation. The synapse results in reaction to the excitation or inhibition depends on type of receptors and neurotransmitters at the synapse.

MODULATION OF SYNAPTIC TRANSMISSION

Pre synaptic inhibition decrease release of neurotransmitter despite an action potential in presynaptic neuron. In an axoaxonic synapse, release of neurotransmitter from one neuron to another neuron results in partial depolarisation of the presynaptic neuron due to less number of calcium (Ca^+) ions entering the presynaptic membrane. The reverse effect occurs upon hyperpolarising release of the neurotransmitter at the axoaxonic synapse leading to rise in neurotransmitter release or presynaptic enabling. The neurotransmitter blocks at times its own nerve cell due to its binding to postsynaptic receptors and also to presynaptic receptors which are called as autoreceptor. These help in feedback loop in signal transduction.

NON-SYNAPTIC CHEMICAL COMMUNICATION

Neuromodulators are other mode of communication substance which are released by neurons and mobile larger distances and are discrete more widely around the nervous system. Neuromodulators are peptides, having chains of amino acids linked by peptide bonds. Neuromodulators affect general behavioural conditions such as observance and anxiety. Hormones are third mode of communication of information within our bodies. Hormones are chemical substances secreted by cells of the endocrine glands or cells in other organs such as the stomach and kidneys. The hormones are described in three categories: amino acid derivatives, peptides and proteins, and steroids. Hormones are secreted by the endocrine gland into the extracellular fluid and are circulated in bloodstream. There are various neurons possessing hormone receptors, affecting our behaviour. For Example; Testosterone is a sex hormone. It increases the aggressiveness of male animals. The following are the glands of the endocrine system:

1. **Adrenal Glands:** Release hormones for stress.
2. **Hypothalamus:** Controls sleep, hunger and appetite.
3. **Ovaries and Testicles:** Controls female and male characteristics.
4. **Pancreas:** releases insulin, controls the levels of glucose present in the body.
5. **Parathyroid Glands:** Retain calcium levels in the body.
6. **Pineal Body:** Maintains sleep cycles and routine biological cycles.

7. **Pituitary Gland:** Maintain endocrine glands.
8. **Thymus Gland:** Controls in the body's immune system.
9. **Thyroid Gland:** Maintains metabolism of the body, including bone growth and heat production.

POSTSYNAPTIC RECEPTORS AND RECEPTOR TYPES

The effect of neurotransmitters on the postsynaptic membrane is opening of various ion channels through ionotropic receptors or channel-linked receptors. The neurotransmitter alone transfers information. There is another type of receptors where ion channels open indirectly calls as metabotropic receptors or G-protein-linked receptors. Upon binding of neurotransmitter to a metabotropic receptor, one more protein gets activated in the cell membrane. This is referred to as a G protein. The G protein activates another chemical known as a second messenger. This opens the ion channels in the postsynaptic membrane. One example is cyclic adenosine monophosphate (cyclicAMP), which is synthesised by adenosine triphosphate (ATP). It is essential to the memory process (Jackson & Ramaswami, 2003) and in the sedative effects of ethanol (Misra & Pandey, 2003).

NEUROTRANSMITTERS

Neurotransmitters are divided into two categories the classical, small molecule and the larger neuropeptide neurotransmitters. Neurotransmitters work in two ways: direct and indirect. In *direct action*, the neurotransmitter opens ion channels by binding to a channel-linked receptor resulting in fast responses. Examples are amino acids and AChE. In *indirect action*, it acts through a G-protein-linked receptor and upon binding of a neurotransmitter, a cascade of intracellular second messenger systems are released and has long-lasting effects. Examples are soluble gases such as NO, and neuropeptides.

The main types of neurotransmitter are described in the following sections

1. **GABA:** GABA is the major blocker among neurotransmitter in the CNS, occurring in 30–40% of all synapses. GABA is present in more concentrations in the substantia nigra and globus pallidus nuclei of the basal ganglia, the hypothalamus, the periaqueductal grey matter and the hippocampus. The concentration of GABA is 200–1000 times more than that of the monoamines or acetylcholine in brain. Various disorders like seizures (Pearl, 2004) and anxiety disorders (Froestl, 2010) occur due to deficiency of GABA.

2. **Glycine:** Glycine is present in the spinal cord, brain stem and retina and occurs only in vertebrates (Hernandes & Troncone, 2009). It is a simple amino acid in both structure and function. Glycine is an inhibitory neurotransmitter. This disinhibition and subsequent spinal hyperexcitability strychnine become poisonous.
3. **Glutamate (Glutamic Acid):** Glutamate is the excitatory neurotransmitter. In the presynaptic membrane, the excitatory act of glutamate is blocked by a chloride-independent membrane transport system that reabsorbs glutamate and aspartate. Processes like learning and memory are formed.
4. **Aspartate:** Aspartate is an excitatory neurotransmitter, exciting NMDA receptors (Grem, King, O'Dwyer, & Leyland-Jones, 1988). Aspartate is primarily in the ventral spinal cord and is inactivated when reabsorbed into the presynaptic membrane.

Following neurotransmitters were called as MONOAMINES

1. **Serotonin:** Serotonin is also known to as 5-hydroxytryptamine (5-HT). Serotonin plays an important role in the central nervous system in managing anger, aggression, mood, sleep, body temperature, sexuality, vomiting and appetite (Folk & Long, 1988). It is found in the cells in the raphe nuclei which are a cluster of nuclei found in the brain stem. Some of the serotonin reuptake inhibitors (SSRIs) and tricyclic antidepressants(TCAs) treat depression. Amphetamine, cocaine and 3,4-Methylenedioxymethamphetamine or MDMA (ecstasy) could block the reuptake of 5-HT.
2. **Melatonin:** Melatonin is derivative from serotonin within the pineal gland and the retina. Melatonin is important in the sleep/wake cycle and is synthesised and secreted during the dark hours and are conserved at a low level during daylight hours. It blocks the synthesis and secretion of other neurotransmitters such as dopamine and GABA.
3. **Noradrenaline:** Noradrenergic neurons are present in locus coeruleus, the pons and the reticular formation in the brain, and project to the cerebral cortex, midbrain and hippocampus. Noradrenaline is an excitatory neurotransmitter, important role in attention and arousal, heart rate, blood pressure and digestive processes.
4. **Dopamine:** Dopamine comes under category of monoamine neurotransmitters. They are present in four dopamine pathways in the brain: the nigrostriatal, the mesocorticolimbic, the mesocortical and the tuberoinfundibular pathways. Dopamine as a hormone blocks the release of prolactin from the anteriorlobe of the pituitary. A shortage of cells producing dopamine leads to the development of Parkinson's disease, while extreme dopamine may be involved in the

development of schizophrenia (Stone, Morrison, & Pilowsky, 2007). In the Parkinsonian patients, L-DOPA(levodopa), the precursor of dopamine, is given as it can cross the blood-brain barrier and go into the CNS.

5. **Acetylcholine:** Acetylcholine (ACh) was the first neurotransmitter to be recognized. It is present in the PNS and the CNS. Ach is an antidepressant where hindering cholinergic receptors and other anticholinergic activity cause dry mouth. ACh is one of the causes of Alzheimer's disease, which have acetylcholine deficiency. Nicotine, in cigarettes, excites nicotinic receptors, while muscarine, a poison in mushrooms, triggers muscarinic receptors.

NEUROPEPTIDES AND NEUROMODULATORS

Neuropeptides, a short chain of amino acid are protein-like molecules which act as neurotransmitters. Postsynaptic cell, but rise or decrease the discharge of neurotransmitter or adapt the reaction of postsynaptic cells.

SUMMARY

In this chapter nervous system has been divided into two categories, the peripheral nervous system (PNS) and the central nervous system (CNS). The nervous system has several cells and these cells contain all the components that make the cell to function. Neurons are further classified on basis of their shape, function or location, or the type of neurotransmitter synthesised and released by them. Neurons have an inward cell having branches of a tree called dendrites and another part emerging from the cell called an axon. The terminal button is at the distal end of a neuron and it places the synaptic vesicles containing neurotransmitter which communicates between the neuron and synapse. The synapse contains the presynaptic region, the synaptic cleft and the postsynaptic region. A concept of non-synaptic chemical communication and the endocrine system is also given in detail. The collections of glands that produce hormones maintain body's growth, metabolism and sexual development and function. The hormones are released into the bloodstream and further moved to target cells and organs in the body. Half of the volumes of CNS are neurons, half consist of supporting cells called glial cells which regulate homoeostasis and provide support and nutrition. Resting potential is -70 mev a cellular membrane and the voltage across membrane when neuron is at rest and how polarisation and depolarisation occurs in the cellular membrane. The neurotransmitters in the nervous system which are released from the end of the terminal button There are different types of neurotransmitter substance and they assist the communication between neuron and target cell and how these transmitters and hormones are important for various functions of body.

QUESTIONS

1. How does synaptic transmission between neurons take place?
2. What is a chemical synapse?
3. How do neurons transmit signals?
4. What happens next after an action potential arrives at a synaptic knob opening calcium channels and allowing calcium to enter the cell?
5. How is an action potential transmitted from neuron to neuron?
6. What is the structure and function of a nerve cell?
7. What is Neuron write the structure and function of neuron?

REFERENCES

Bennett, M. V., & Zukin, R. S. (2004). Electrical coupling and neuronal synchronization in the mammalian brain. *Neuron, 41*(4), 495–511. doi:10.1016/S0896-6273(04)00043-1 PMID:14980200

Christiane, S. H., Hiroshi, M., & Mario, M. (2017). *GABA and Glutamate: Their Transmitter Role in the CNS and Pancreatic Islets, GABA And Glutamate - New Developments In Neurotransmission Research*. IntechOpen. Available from: https://www.intechopen.com/books/gaba-and-glutamate-new-developments-in-neurotransmission-research/gaba-and-glutamate-their-transmitter-role-in-the-cns-and-pancreatic-islets

Fields, R. D., & Stevens-Graham, B. (2002). New insights into neuron-glia communication. *Science, 298*(5593), 556–562. doi:10.1126cience.298.5593.556 PMID:12386325

Folk, G. E. Jr, & Long, J. P. (1988). Serotonin as a neurotransmitter: a review. *Comparative Biochemistry and Physiology Part C Toxicology & Pharmacology, 91*(1), 251-7.

Frank, W. P. (2009). Roles of glial cells in synapse development. *Cellular and Molecular Life Sciences, 66*(13), 2037–2047. doi:10.100700018-009-0005-7 PMID:19308323

Grem, J.L,, King, S.A., O'Dwyer, P.J., & Leyland-Jones, B. (1988). Biochemistry and clinical activity of N-(phosphonacetyl)-L-aspartate: a review. *Cancer Research, 48*(16), 4441-54.

Hernandes, M. S., & Troncone, L. R. (2009). Glycine as a neurotransmitter in the forebrain: A short review. *Journal of Neural Transmission (Vienna, Austria), 116*(12), 1551–1560. doi:10.100700702-009-0326-6 PMID:19826900

Jackson, T., & Ramaswami, M. (2003). Prospects of memory-modifying drugs that target the CREB pathway. *Current Opinion in Drug Discovery & Development, 6*, 712–719. PMID:14579521

Jewett, B. E., & Sharma, S. (2019). *Physiology, GABA*. Treasure Island, FL: Stat Pearls Publishing.

Kettenmann, H., & Verkhratsky, A. (2016). Advances of Experimental Medicine Biology. *Springer Nature (Switzerland), 949*, 1–24.

Li, Y., Sun., H., Chen, Z., Xu, H., Bu, G., & Zheng, H. (2016). Implications of GABAergic Neurotransmission in Alzheimer's Disease. *Frontiers in Aging Neuroscience, 23*(8), 31. doi:10.3389/fnagi.2016.00031

Lodish, H., Berk, A., Zipursky, S. L., Matsudaira, P., Baltimore, D., & Darnell, J. W. H. (2002). Molecular Cell Biology. Science.

Misra, K., & Pandey, S. C. (2003). Differences in basal levels of CREB and NPY in nucleus accumbens regions between C57BL/6 and DBA/2 mice differing in inborn alcohol drinking behavior. *Journal of Neuroscience Research, 74*(6), 967–975. doi:10.1002/jnr.10831 PMID:14648603

Pearl, P. L., & Gibson, K. M. (2004). Clinical aspects of the disorders of GABA metabolism in children. *Current Opinion in Neurology, 17*(2), 107–113. doi:10.1097/00019052-200404000-00005 PMID:15021235

Pfrieger, F. W. (2010). Role of glial cells in the formation and maintenance of synapses. *Brain Research. Brain Research Reviews, 63*(1-2), 39–46. doi:10.1016/j.brainresrev.2009.11.002 PMID:19931561

Terni, B., López-Murcia, F.J., & Llobet, A. (2011). Neuroglia living nerve glue. *Fortschrtte der Neurologie Psychiatrie, 79*(10), 588-97.

Terni, B., López-Murcia, F. J., & Llobet, A. (2016). Role of neuron-glia interactions in developmental synapse elimination. *Brain Research Bulletin, 129*, 74–81. doi:10.1016/j.brainresbull.2016.08.017 PMID:27601093 von Bernhardi, R., Eugenín-von Bernhardi, J., Flores, B., & Eugenín León, J. (2016). Glial Cells and Integrity of the Nervous System. *Advances of Experimental Medicine*, *949*, 1–24. doi:10.1007/978-3-319-40764-7_1 PMID:27714682

KEY TERMS AND DEFINITIONS

Cells: Cells are the elementary building slabs of living things. The human body is composed of trillions of cells, all with their own focussed function. Cells are the basic structures of all living organisms. Cells deliver structure for the body, take in nutrients from food and carry out significant function.

GABA: Gamma-aminobutyric acid, or γ-aminobutyric acid, or GABA, is the principal inhibitory neurotransmitter in the developmentally mature mammalian central nervous system. Its prime role is reducing neuronal excitability throughout the nervous system. GABA is sold as a dietary supplement.

Neuron: A neuron (also called neurone or nerve cell) is a cell that transmits electrical impulses. Neurons are the elementary (functional and structural) units of our nervous system. Every neuron is made of a cell body (also called soma or cyton), dendrites and an axon. Dendrites and axons are nerve fibers.

Neurotransmitters: Neurotransmitters are endogenous chemicals that allow neurotransmission. It is a type of chemical messenger which transfers signals crosswise a chemical synapse, such as a neuromuscular junction, from one neuron (nerve cell) to another target neuron, muscle cell, or gland cell.

Chapter 4
Biochemistry of the Brain

Sumit Kumar
Jawaharlal Nehru University, India

ABSTRACT

Biochemistry is the cornerstone of biological science, feeding the new discoveries and providing us essential and fundamental discoveries vital to understand natural science. Biochemistry has been long recognized as a central pillar in neurochemistry, especially the imbalance of the enzyme that leads to the faulty functioning of the brain. Furthermore, we had also comprehensively discussed the metabolic and genetic aspect of major mental disorders and the genetics of intellectual disability and behavioral disorders.

CHAPTER OUTLINE

- Introduction
- Biochemical, metabolic and genetic aspect of major mental disorders,
- The genetics of Intellectual disability and behavioral disorders
- Other behavioral disorders
- Conclusion

Learning Objectives

- Describing the biochemical, aspect of mental disorders, mental retardation and, behavioral disorders
- To describe the role of metabolic system in psychiatric disorders

DOI: 10.4018/978-1-7998-2860-0.ch004

INTRODUCTION

The biochemistry is the cornerstone in the biological science, feeding the new discoveries and providing us essential and fundamental discoveries vital to understand natural science. Biochemistry has been long recognized as central pillar in neurochemistry, especially the imbalance of the enzyme, which leads to the faulty functioning of the brain. Understanding the biochemical deficiency is also vital for rectification of the metabolic error, and restoringthe normal functioning of the brain.

The neurochemistry was emerged within neuroscience during 1940s, when Hofmann discovered the hallucinogenic properties of synthetic ergot derivative, and lysergic acid diethylamide (LSD) in 1943 (Hofmann., 1970). The creation of vivid experience by the above compounds forced him to think that if alkaloid can cause mental imbalance, then a toxin may create a natural mental disease. He hypothesized that psychiatric disorders are linked to specific toxic metabolites and gave "the toxic metabolite hypothesis".Hoffer, Osmond and Smythies together in 1954 proposed "adrenochrome theory", which based on the idea that schizophrenia is caused by adrenochrome. Later it got support from Saskatchewan group, who reported the hallucinogenic action of adrenochrome in man. Although all theories have been disproved so far, but it gave the basic foundation to neurochemistry. DNA represents the central point in governing all biochemical processes in the body. The emergence of genome biology, DNA sequencing, informatics, and genome manipulation gave further boost to trace and find out the source of brain disorder easily. Therefore the central theme of the present chapter is the role of different genetic deficiencies in deregulation of different biochemical processes and then how it contributes to the psychiatric disorder.

BIOCHEMICAL AND METABOLIC ASPECT OF MAJOR MENTAL DISORDERS

Psychotic disorders are always traced to some metabolic deficiencies. Stanley Mann working at Maudsley Hospital, London was the first person to report higher chance of deregulated carbohydrate metabolism or consistence hyperglycemia in mentally ill patients and thus he linked the chronic mental cases with the disordered carbohydrate metabolism (Sourkes., 1961). Since then, over four hundred types of mental disorders are identified and placed into five major categories as follow:

1. Disorders diagnosed during childhood.
2. Organic mental disorders.
3. Substance-related disorders.

4. Mood disorders and personality disorders.
5. Schizophrenia and other psychotic disorders.

DISORDERS DIAGNOSED DURING CHILDHOOD

Childhood disorders are due to severe errors in the certain metabolic pathway. It causes discrepancies in the way child typically learn, behave, or handle their distress and emotions. These are mostly recurrent psychotic attacks, emotional/chronic delusion, mood extremes, and sometime may cause delusions (Morgan et al., 2016). The most common diseases recognized during childhood are homocystinuria, Wilson disease, lysosomal storage disorders, adrenoleukodystrophy, and so on. These disorders cause mild to severe mental retardation and may also cause some personality change. These mental retardations are mostly due to the deregulation in the phenylketonuria, hyperglycinemia, creatine deficiency syndrome, homocystinurias, nonketotichyperglycinemia, monoamine oxidase-a-deficiency, succinic semialdehyde dehydrogenase deficiency, creatine transporter defect, and α/β-mannosidosis.The urea cycle is one of the pathways that cause mental retardation. Urea is a toxic substance produced in the body, and thus elimination of the urea is almost necessary for the basic functioning of the brain. The diseases of the urea cycle include a group of six distinct enzyme deficiencies that cause deregulation in the urea cycle and also rise in the serum ammonia level. The disease is detected by the abnormalities in the blood amino acids, namely glutamine, ornithine, citrulline, and arginine. Ornithine transcarbamylase deficiency is one of the most general things in deregulation of urea excretion.

ORGANIC MENTAL DISORDERS

The organic mental disorder may be permanent or temporary arising mainly due to disturbance, abnormalities in the brain, hormonal deregulation, toxic compound exposure, liver disease, kidney disease, and metabolic deregulation that cause dysfunction in the brain. The disease like delirium causes serious mental dysfunction in affected person may run from 1-6 hours. It causes sudden disability in the brain leading to confused thinking and reduced awareness about the environment. Organic mental disorders are may be due to changes in metabolic balance (such as low glucose or sodium), dehydration, medication, infection, surgery, constipation or alcohol or drug intoxication, etc. The biochemical basis of the organic mental disorder is mainly due to the accumulation of toxic chemicals or lack of energy that seriously hamper the normal functioning of the brain.

SUBSTANCE-RELATED DISORDERS

Substance-related disorder is due to addictive drugs that activate the brain's reward system. Avoidance or withdrawal causes serious dysfunction in the system (Kosten et al., 2018). The pharmacological and biochemical basis of the substance-related disorders lay in excitation or suppression of neurons. These drugs are mainly divided into ten classes namely alcohol, caffeine, cannabis, hallucinogens like LSD, phencyclidine, psilocybin, inhalants like volatile hydrocarbons especially paint thinner, certain glues, opioids like fentanyl, morphine, oxycodone, sedatives, hypnotics, anxiolytics like lorazepam, secobarbital, stimulants like amphetamines, cocaine, tobacco and many other drugs especially anabolic steroids.

Alcohol: It depresses central nervous system. The common symptoms include central nervous system dysfunction, craniofacial malformations, and inhibited growth development. The behavioral effects include attention deficit, distractibility; impulsivity; restlessness; inept social interactions; self-control problems; memory loss, problem-solving, and organizational difficulties.

Marijuana: It produces relaxation and euphoria. Opiates produce pain reduction, constipation and powerful euphoria (Pacula et al., 2015). Hallucinogens produce sensory, perceptual, and cognitive hallucinations. Sedatives part depress the central nervous system and induce sleep. While stimulants part increase activity of the central nervous system, decrease fatigue, increase alertness and arousal, and can include euphoria. The wide arrays of behavioral effects of marijuana have been hypothesized, including social, learning, language and attention deficits.

Lead Poisoning: It permanently and progressively damages to the central nervous system. It may lead to encephalitis developmental and outcomes may include seizures, cerebral palsy, and retardation.

Microcephaly: It is transmitted genetically as an autosomal recessive trait and most often the result of prenatal toxin exposure. It is associated with severe intellectual disabilities and an energetic, outgoing personality.

Hydrocephalus: It results from polygenic inheritance or prenatal exposure to toxins.

Metabolic aspect of mental disorders, mental retardation and behavioral disorders:

Peoples with mental deficit such as schizophrenia and bipolar disorder have an increased prevalence of metabolic syndrome and thus metabolic syndrome act as causative factor in mental retardation.

Nutritional Deficiencies: It can occur pre- or post-natally. It can lead to as much as a 40% deficit in brain cells. The effects of early malnutrition,especially during gestation or the first six months of life,have long been viewed as irreversible.

Phenylketonuria: Phenylketonuria is an inherited disorder that increases the levels of a substance called phenylalanine in the blood. It is due to the reduction in metabolism of the amino acid phenylalanine. Untreated, phenylketonuria may result in intellectual disability, seizures, behavioral problems, and mental disorders.

MOOD DISORDERS AND PERSONALITY DISORDERS

Mood disorders are formed by long-lasting disturbances in emotions that strongly affect feelings, thoughts, and behaviors. It divided into two categories, namely major depressive disorder and bipolar spectrum disorder. The major depressive disorder cause clinical depression and may last about two weeks. While bipolar spectrum disorder cause alternating major depressive and manic episodes with alternative periods of normal brain functioning.

SCHIZOPHRENIA AND OTHER PSYCHOTIC DISORDERS

Schizophrenia is a severe mental disorder in which thinking and emotion are so impaired that the individual is seriously out of contact with reality. The enlarged ventricles (i.e., spaces) in the brain and/or decreased volume in frontal & temporal lobes and relatively low glucose metabolism in frontal areas indicate deterioration or atrophy of brain tissue. Schizophrenia arises due to excess dopamine activity at certain synaptic sites. The treatment of schizophrenia includes phenothiazines that reduce the dopamine activity and so psychotic symptoms.

The major mental health conditions are namely schizophrenia, bipolar disorder, major depressive disorder, autism spectrum disorder, attention deficit hyperactivity disorder.

Schizophrenia: Schizophrenic is mental disorders characterized by fundamental and characteristic distortion of thinking, perception, and affects that are inappropriate or blunted (Guloksuz et al., 2018). Clear consciousness and intellectual capacity are usually maintained although certain cognitive deficits may evolve during time course. Peoples with schizophrenia have an increased prevalence of metabolic syndrome and thus metabolic syndrome act as causative factor in mental retardation.

Bipolar Disorder: Characteristics of a bipolar disorder are distinct period of abnormally and persistently elevated, expansive or irritable mood. It has some key genetic elements like ANK3, CACNA1C, and CLOCK genes, which are known to be responsible for bipolar disorder (Bengesser et al., 2019). During the period of mood disturbance, at least three of the following symptoms have persisted (four if the mood is only irritable) and have been persistent to a significant degree: inflated self-esteem or grandiosity, decreased need for sleep, more talkative than usual or pressure to keep talking; Flight of ideas or subjective experience that thoughts are racing. The affected individual also have high degree of distractibility, i.e. attention too easily drawn to unimportant or irrelevant external stimuli and also excessive involvement in pleasurable activities which have a high potential for painful consequences, e.g. unrestrained buying sprees, sexual indiscretions, or foolish business investments. It is caused by toxins, drugs abuse, stimulants (amphetamines, cocaine), hallucinogens (LCD, PCP), prescription medications, antidepressants, L-dopa, corticosteroids, neurologic nondominant frontal CVA, nondominant frontal tumors, Huntington's disease, multiple sclerosis, neurosyphilis, HIV, hyperthyroidism, and cushing's disease etc. Peoples with bipolar disorder have an increased prevalence of metabolic syndrome and thus metabolic syndrome act as causative factor in mental retardation.

Major Depressive Disorder: Major depressive disorder is one of the most common major depressive disorder. It causes significant medical condition that interferes with many daily life tasks. It affects the mood and behavior as well as various physical bodily functions, such as appetite, sleep and breathing. Peoples with it lose interest in activities once they like and also give trouble performance. Major depressive disorder cause clinical depression that lasts about two weeks,

Autism Spectrum Disorder: These children are unable to mentally process information from the body language, gestures, facial expression or tone of voice of other people (Kincaid et al., 2017). The most common risk factors associated with autism are certain infections, such as measles, rubella, herpes simplex virus, mycoplasma, *Chlamydia pneumoniae*, *Borreliaburgdorferi*, toxins including valproic acid, alcohol, cocaine, pesticides and air pollution, fetal growth restriction, certain human leukocyte antigen genes/haplotypes and autoimmune diseases leading to Infection, inflammation, and immune activation during pregnancy (Abib et al., 2018). Certain immunological abnormalities and above infection invoke cytokines response namely interleukin [IL]-1R, IL-6, IL-8, IL-10, and IL-17, immunoglobulins formation, inflammation, and other immune cellular activation result in development of autism (Abib et al., 2018). However multiple studies disproved the claim and thus autism vaccine is hang in limbo.

Attention Deficit Hyperactivity Disorder: There are mainly three forms of attention-deficit/hyperactive disorder (Hechtman., 2019). It occurs due to the interaction of environmental factors with certain genetic predispositions. The common risk factors of attention deficit hyperactive disorder are causative factors responsible for schizophrenia and schizophrenia itself. Stanley and his coworkers have shown the role of brain developmental and biochemical factors in development of attention-deficit/hyperactive disorder. They found that regional-specific alterations in precursors of membrane phospholipids in attention deficit/hyperactive disorder kids. It results in underdevelopment of neuropil, a network of unmyelinatedaxons,dendrites and glia – in the basal ganglia region. Children affected with attention-deficit/hyperactive disorder also lack neuropil expansion in prefrontal cortex. This indicates that membrane phospholipids lead neuronal developmental alterations in prefrontal cortex and basal ganglion is responsible for attention deficit/hyperactive disorder. It is categorized into three types as follow:

Attention deficit disorder: Attention deficit disorder is the inability to pay attention.

Hyperactive disorder: Hyperactivity disorder is being hyperactive and impulsive.

Attention deficit hyperactive disorder:Attention deficit hyperactive disorder is a combination of the attention deficit and hyperactive disorder.

THE GENETICS OF MENTAL RETARDATION (INTELLECTUAL DISABILITY)

The brain is the most complex piece present in the biological system. The high level of complexity arises due to gene–gene and gene-environment interactions (Kumar, Singh,&Meena, 2016). The gene is basic biological units carrying inherited physical, mental, or personality traits (Vissers,Gilissen,&Veltman, 2016). Approximately 30,000 genes are present in each human cell, occupying specific positions on chromosomes. Pairs of genes carrying the same trait are called homozygous while pairs of genes carrying different traits are called heterozygous condition. The specific gene-gene interaction produces a special phenotype in form of neuronal interconnection that further interacts with the billions of neuron to produce a complex outcome known as behavior. This high level complexity has made the advancement in neuroscience and genetics of mentalillness exceedingly difficult, around 10-50% of cases of mild retardation are related to genetic etiologies.

The genes are arranged in chromosomes, a threadlike or rod like bodies that is composed of genes containing genetic information. Each human cell contains 23 pairs of chromosomes; one member of each pair is donated from each parent. The autosomes matching pairs constitute 22 of the 23 pairs. Sex chromosomes are the

23rd pair responsible for the determination of gender. Mother always donates an X chromosome, while Father donatesingle X orY chromosome. Thus XX represent female gender and XY represent male gender.

The gene is either dominant or recessive.

Dominant Genes Disorder: It means abnormal gene from one parents can cause disorder. Several rare physical disorders like marfan's syndrome, dwarfism, and neurofibromatosis type one are carried as dominant traits, and these are often structural disorders.

Recessive Genes Disorder: It cannot control their partners, and will not express when paired with a dominant gene. Taysach's disease is recessive inherited and disproportionately prevalent among Ashkenazi Jews. It is typically manifested late in the child's 1st year and cause severe retardation, convulsions, blindness, paralysis, and death by the age of 4. Phenylketonuria is another recessive disorder.

Phenylketonuria: It is most common of the genetic disorders causing aggressiveness, hyperactivity, and retardation. It is most publicized success story in the literature. It is virtually eliminated as a causative factor in severe retardation. The treatment regimen for PKU is related to restrictions in intake of phenylalanine, common in high-protein foods.

Sex-Linked Inheritance in Mental Retardation: It is also called X-linked inheritance. It occurs through recessive traits carried on the X chromosome. Females are affected only if both parents pass along the recessive gene, while males are affected if the mother passes along the recessive gene, because there is not another X chromosome. Around 200 genes on X-chromosome linked to more than 70 disabilities intellectual disability like Lesch-Nyhan disease, fragile X syndrome and so on. The 50 genes have been mapped.

Lesch-Nyhan Disease: It is second most common metabolic disorder associated with extreme impulse to self-injure and, to a lesser extent, injure others. Biting is the most prominent form, but other forms of self-injury are common. Some form of restraint is usually necessary for physical and psychological safety.

Fragile X Syndrome: One in 260 women is carriers of fragile X syndrome. The fragile X syndrome transmitted from mother to son, which can be pre-mutated (carrier) to mutated (affected) or mutated (affected) to mutated (affected) or pre-mutated (carrier) to pre-mutated (carrier). The father can be pre-mutated (unaffected and a carrier) and pass the fragile X gene on to his daughter (Hodapp et al., 2016). It is not transferred by male-to-male transmission. Fragile X occurs in about 1 in 1,500 males and about 1 in 1,000 females in the general population. Mental Retardation is most common in fragile X syndrome. The affected individual show gaze avoidance, attention problems, repetitive speech (including echolalia and palilalia) and repetitive behaviors. Fragile X mutations may increase the risk of a child developing autistic-like tendencies. It delayed the development of speech and language by age 2.

Down Syndrome: Down syndrome prevalence is 1 in 700-1000 births. Approximately 5-7% of people with mental retardation have Down syndrome. It is caused by trisomy of chromosome number 21. The risk factors of Down syndrome are old age mothers (1 in 30 births to women over age 45), lack of balanced diet, overweight and so on. It causes severe mental retardation, physical growth delay and characteristic face. Two genes implicated in Alzheimer's disease are located on chromosome, thus it also increased the risk for developing Alzheimer's disease.

Prader-Willi Syndrome: It is caused by a deletion of the paternal chromosome number 15. The affected individual show moderate mental retardation and significant externalizing behaviors like tantrums, aggression especially passive aggression and so on.

Williams Syndrome: The affected individual show a wide variation in IQ, but most have mild mental retardation to borderline intellectual functioning. Williams Syndrome causes the developmental delays in speech/language, motor and academic skills are very common. Limited visual-spatial development, poor visual-motor skills (e.g., handwriting problems), mathematical and numerical knowledge and musical abilities are other common future.

Klinefelter Syndrome: The affected male individual receives extra X chromosome(s). Psychosocial problems include language problems, difficulties in social relationships, mood disturbance and intellectual deficits increase with number of X chromosomes.

Turner Syndrome: It is results from the absence of one of the X chromosomes (XO) in females. It causes only learning disability rather than mental retardation. Other common problems are in spatial relations, mathematics, memory, attention, and social competence.

OTHER BEHAVIOURAL DISORDERS

1. Generalized anxiety disorder
2. Substance abuse disorder
3. Eating disorder
4. Learning disorder
5. Conduct disorder
6. Comorbidity of behavioral disorders

Generalized anxiety disorder: Generalized anxiety disorder is characterized by persistent and excessive worry and exaggerated anxiety about general life events. The individual suffering from generalized anxiety disorder always anticipate disaster and may not stop thinking about health, friends, money, responsibilities, future, family, work, or college. The consistence unrealistic fear affects their work activities, working performance, relationship with coworkers, friends, and family members.

Substance Abuse Disorder: Substance abuse disorder is a brain disease associated with changes in the brain. It occurs due to misuse and abuse of certain drugs and psychotic substances. The person addicted to drugs behaves abnormally, slow responsive, lacking willpower, and become hopeless. Over the time addiction causes changes in the brain, namely prefrontal cortex, and limbic system. It mainly affect the neural connections associated with dopamine pathway, seratonin pathway, and brain chemicals, especially dopamine, serotonin, endorphin, glutamate.

Eating Disorder: Eating disorders are two types namely anorexia nervosa, and bulimia nervous (Linardon et al., 2016). In anorexia nervosa, individual refuse to maintain standard body weight and so always above to normal weight and height. It cause disturbance in the way in which one's body weight or shape is experienced. While in bulimia nervous, the body weight is reduced due to undue influence of body weight on self-evaluated shape, It cause hormonal disbalance in female, resulting in amenorrhea. The affected individual may also relay in controlling weight by restricting food intake, self-induce vomiting, laxatives intake or diuretics (purging behaviors), exercises excessively, appetite suppressants intake.

Learning Disorder: Learning disorder is the common name for something that affects a person's ability to interpret what they see or hear. Learning disorder can be divided into three main categories namely developmental speech and language disorders, academic skills disorders and "Other," a catch-all that includes certain coordination disorders and learning handicaps not covered by the other terms, It is caused by errors in fetal brain development, genetic factors, problems during pregnancy or delivery, tobacco, alcohol, and other drug use.

Toxins in the Child's Environment: Brain development begins within a week of conception. By the time a baby is born, about 100 billion neurons or brain cells have been produced. Many of them already migrate to their final destinations and begin forming synaptic connections. Much of the connecting strengthening of neurons in newborn baby are influenced by the environment. Toxic such as medications, illicit drugs like opiates, cocaine, smoking, environmental toxins like mercury from fish, PCBs, aspartame and radiation etc. cause serious dysfunction in exposed children brain.

Conduct Disorder: The term conduct disorder has traditionally been used to characterize children who display a broad range of behaviors that bring them into conflict with their environment. These include behaviors that are probably best described as coercive or oppositional, temper tantrums, defiance, and noncompliance. It also includes behaviors of a more serious nature such as cruelty to people or animals, aggressiveness, stealing etc.

Comorbidity of Behavioral Disorders: Comorbidity is defined as a situation in which an individual, who has been diagnosed with one specific disorder, is also found to meet diagnostic criteria for one or more additional disorders (Karaca et al., 2017). Blashfield has referred to comorbidity as the co-occurrence of different diseases in the same individual. Likewise, Caron and Rutter has defined comorbidity as the simultaneous occurrence of two or more unrelated conditions. The multiple genetic and epigenetic factor, environmental factor and stress has been recognized as causative factor in comorbidity disorder. Hypothalamic-pituitary-adrenal (HPA) axis hyperactivity increase vulnerability to substance use disorders through dopamine pathway.

CONCLUSION

The biochemical, metabolic and genetic factors are known to play a major role in shaping mental status and behaviors. In the present chapter I comprehensively discussed the causative factors of brain disorder and offer new light on the pathophysiology of brain disorder.

Questions

1. Explain the mental disorder.
2. Explain the mental retardation.
3. Explain various factors responsible for autism.
4. What is bipolar disorder?
5. Discuss the gene and chromosomal abnormalities associated with mental retardation.

REFERENCES

Abib, R. T., Gaman, A., Dargél, A. A., Tamouza, R., Kapczinski, F., Gottfried, C., & Leboyer, M. (2018). Intracellular Pathogen Infections and Immune Response in Autism. *Neuroimmunomodulation, 25*(5-6), 271–279. doi:10.1159/000491821 PMID:30130799

Bengesser, S. A., Mörkl, S., Painold, A., Dalkner, N., Birner, A., Fellendorf, F. T., ... Reininghaus, E. Z. (2019). Epigenetics of the molecular clock and bacterial diversity in bipolar disorder. *Psychoneuroendocrinology, 1*(101), 160–166. doi:10.1016/j.psyneuen.2018.11.009 PMID:30465968

Guloksuz, S., & Van Os, J. (2018). The slow death of the concept of schizophrenia and the painful birth of the psychosis spectrum. *Psychological Medicine, 48*(2), 229–244. doi:10.1017/S0033291717001775 PMID:28689498

Hechtman, L. (2019). ADHD medication treatment and risk of psychosis. *The Lancet. Psychiatry, 6*(8), 632–633. doi:10.1016/S2215-0366(19)30248-2 PMID:31221558

Hodapp, R. M., Dankner, N. A., & Dykens, E. M. (2016). Behavioural phenotypes/genetic syndromes. *Psychiatric and Behavioural Disorders in Intellectual and Developmental Disabilities*, 196.

Hofmann, A. (1970). *The Discovery of LSD and Subsequent Investigations on Naturally Occurring Hallucinogens in Discoveries in Biological Psychiatry*. JB Lippincott Company.

Karaca, S., Saleh, A., Canan, F., & Potenza, M. N. (2017). Comorbidity between behavioral addictions and attention deficit/hyperactivity disorder: A systematic review. *International Journal of Mental Health and Addiction, 15*(3), 701–724. doi:10.100711469-016-9660-8

Kincaid, D. L., Doris, M., Shannon, C., & Mulholland, C. (2017). What is the prevalence of autism spectrum disorder and ASD traits in psychosis? A systematic review. *Psychiatry Research, 250*, 99–105. doi:10.1016/j.psychres.2017.01.017 PMID:28152400

Kosten, T. R., Haile, C. N., Newton, T. F., & De La Garza II, R. (2018). Alcohol and Other Substance Use Disorders. *The American Psychiatric Association Publishing Textbook of Neuropsychiatry and Clinical Neurosciences*, 413.

Kumar, S., Singh, R.K., & Meena, R. (2016). Emerging targets for radioprotection and radiosensitization in radiotherapy. *Tumor Biology, 37*(9), 11589-609.

Linardon, J., Phillipou, A., Newton, R., Fuller-Tyszkiewicz, M., Jenkins, Z., Cistullo, L. L., & Castle, D. (2018). Testing the relative associations of different components of dietary restraint on psychological functioning in anorexia nervosa and bulimia nervosa. *Appetite*, *128*, 1–6. doi:10.1016/j.appet.2018.05.138 PMID:29807122

Morgan, C., & Anderson, G. C. (2016). Childhood adversities and psychosis: Evidence, challenges, implications. *World Psychiatry; Official Journal of the World Psychiatric Association (WPA)*, *15*(2), 93–102. doi:10.1002/wps.20330 PMID:27265690

Pacula, R. L., Powell, D., Heaton, P., & Sevigny, E. L. (2015). Assessing the effects of medical marijuana laws on marijuana use: The devil is in the details. *Journal of Policy Analysis and Management*, *34*(1), 7–31. doi:10.1002/pam.21804 PMID:25558490

Sourkes, T. L. (1961). The biochemistry of mental disease. *Canadian Medical Association Journal*, *85*(9), 487. PMID:20326876

Vissers, L. E., Gilissen, C., & Veltman, J. A. (2016). Genetic studies in intellectual disability and related disorders. *Nature Reviews. Genetics*, *17*(1), 9–18. doi:10.1038/nrg3999 PMID:26503795

KEY TERMS AND DEFINITIONS

Attention Deficit Hyperactive Disorder: It is a combination of the attention deficit, hyperactive and impulsive behavior.

Bipolar Disorder: Bipolar disorder is distinct period of abnormally and persistently elevated, expansive, and irritable mood.

Intellectual Disability: It is also known as general learning disability and mental retardation. It is a generalized neurodevelopmental disorder characterized by significantly impaired *intellectual* and adaptive functioning.

Mental Disorder: A mental disorder, also called a mental illness or psychiatric disorder, is a behavioral or mental pattern that causes significant distress or impairment of personal normal behavior.

Metabolic Deficit: This deficit can be a result of reduced calorie intake, increased calorie expenditure through exercise, or a combination of the two.

Schizophrenia: Is a mental disorder characterized by fundamental and characteristic distortions of thinking and perception and affects that are inappropriate or blunted.

Substance Abuse: Substance abuse disorder is a brain disease associated with changes in the brain. It occurs due to misuse and abuse of certain drugs and psychotic substances.

Chapter 5
Neurobiology of Sensory Motor Systems and Internal Environment

Pradeep Kallollimath
Dharwad Institute of Mental Health and Neurosciences, India

ABSTRACT

Control of motor function—our activities like walking, lifting an object, writing, etc.—is accomplished through integrated and coordinated action of motor neurons. Motor neurons can be classified into two broad categories: upper motor neurons and lower motor neurons. Upper motor neurons have cell bodies in the motor cortex of brain and carry the impulses from cortex to spinal cord segment. Cell bodies of lower motor neurons are located in the anterior horn of spinal cord. Axons of lower motor neuron end in neuromuscular junction, and excitation of muscle fibres leads contraction of muscle. Different types of receptors carry different sensations like touch, pain, position, and vibration through sensory nerve fibres. Fibers mediating fine touch and proprioception ascend in the dorsal columns to the medulla, where they synapse in the gracile and cuneate nuclei. This chapter explores the neurobiology of sensory motor systems and the internal environment.

CHAPTER OUTLINE

- Sensory system
- Motor system
- Regulation of internal environment
- Limbic system

DOI: 10.4018/978-1-7998-2860-0.ch005

- Autonomic system
- Neuroendocrine system

Learning Objectives: To learn basic anatomy, physiology and clinical implications of sensory, motor, autonomic and limbic system.

INTRODUCTION

- **Sensory System:** The sensory system carries sensory information from external and internal environment. Different types of receptors carry different sensations like touch, pain, position and vibration through sensory nerve fibres. Impulses that arise by stimulation of receptors are carried by sensory nerves, and then conveyed through fiber tracts to higher centers in the brain. The sensory system can be classified as somatic and visceral sensation, with general and special varieties of each. General somatic afferent fibers carry touch, pain, position and vibration sense. General visceral afferent fibers carry impulses from internal organs. Special somatic and special visceral afferent fibers mediate special senses- smell, vision, taste, hearing, and vestibular sensation.

  The different sensory modalities are carried over nerve fibers of different size. Sensory impulses are carried to the dorsal (posterior) root ganglia, and then into the spinal cord. After one or more synapses, the impulses ascend through specific fiber tracts to reach sensory areas of the brain. Fine touch, position, and vibration sensations are carried through the posterior column/medial lemniscus system. Pain and temperature senses are carried through spinothalamic tracts. The major sensory pathways are depicted in Figure1.

  Figure 1 depicts the light touch, pressure, position, and vibration pathways from the body and face are indicated by the dashed line; the pain and temperature fibers from the body and face are indicated by the solid line. Fibers from these various sources ultimately converge on the ventral posterior nuclei of the thalamus, which projects via the thalamic radiations to the primary sensory cortex in the postcentral gyrus. V, trigeminal; VPL, ventral posterior lateral; VPM, ventral posterior medial.
- **Sensory Receptors:** There are many different types of receptors in the skin, subcutaneous tissues, muscles and internal organs. Receptors convert various types of sensory information into nerve impulses. Sensory end-organs are found in the skin and mucous membranes throughout the body. They are denser on the tongue, lips, genitalia, and fingertips and farther apart on the upper arms, buttocks, and trunk. Receptor stimulation causes a change in

the permeability of its membrane that gives rise to a local, non propagated potential known as receptor potential. Receptor potentials induce action potentials in the nerve. Receptors may be free nerve endings, ormay be encapsulated or specialized sense organs.

Examples

- **Free Nerve Endings:** Fine, unmyelinated terminal fibers mediate pain (nociceptors). Merkel cell endings (tactile discs or menisci) are specialized nerve endings that function as mechanoreceptors which sense pressure.Other examples of receptors include Golgi tendon organs, muscle spindles, Ruffini's endings, and Meissner's and Pacinian corpuscles.
- **Dermatomes:** Sensory nerve roots supply cutaneous innervation to specific dermatomes. There is significant dermatomal overlap; hence clinical deficit from an isolated root lesion is typically much more restricted than that expected from the anatomical geography of the dermatome.
- **Ascending Tracts in Spinal Cord:** The oval-shaped dorsal root ganglia (DRG) lie on the posterior root in the inter vertebral foramen. The DRG neurons are unipolar. A single process leaves the cell and then devides into peripheral and central branches. The peripheral processes conduct afferent impulses through the dorsal roots toward the cell body. The dorsal root is divided into a medial zone, which contains large fibers carrying proprioceptive(position) sensation and a lateral zone, which contains small fibers carrying pain and temperature sensation. Within the spinal cord (see the figure 2), pain and temperature carrying second-order nerve fibres decussate over several segments by way of the ventral white commissure and ascend as the lateral spinothalamic tract to reach the thalamus (ventral posterolateral nucleus). Fibers carrying tactile sensibility (light touch) also cross over to the opposite side through the ventral white commissure and ascend as the ventral spinothalamic tract to reach the ventral posterolateral nucleus of the thalamus.

  The large myelinated fibers of the medial bundle of the dorsal root pass over the dorsal horn into the dorsal column of the spinal cord. The fibers from the lower part of the body (sacral, lumbar, and lower thoracic) are located more medially and constitute the fasciculus gracilis, whereas those coming from the upper part of the body (upperthoracic and cervical) occupy a more lateral position and constitute the fasciculus cuneatus. The fasciculus gracilis ends in the nucleus gracilis of the medulla; the fasciculus cuneatus terminates in the nucleus cuneatus of medulla. The axons of these two nuclei decussate in the lower medulla and ascend as the medial lemniscus to reach the ventral posterolateral nucleus of the thalamus. These fibers carry information

Figure 1. Sensory pathways

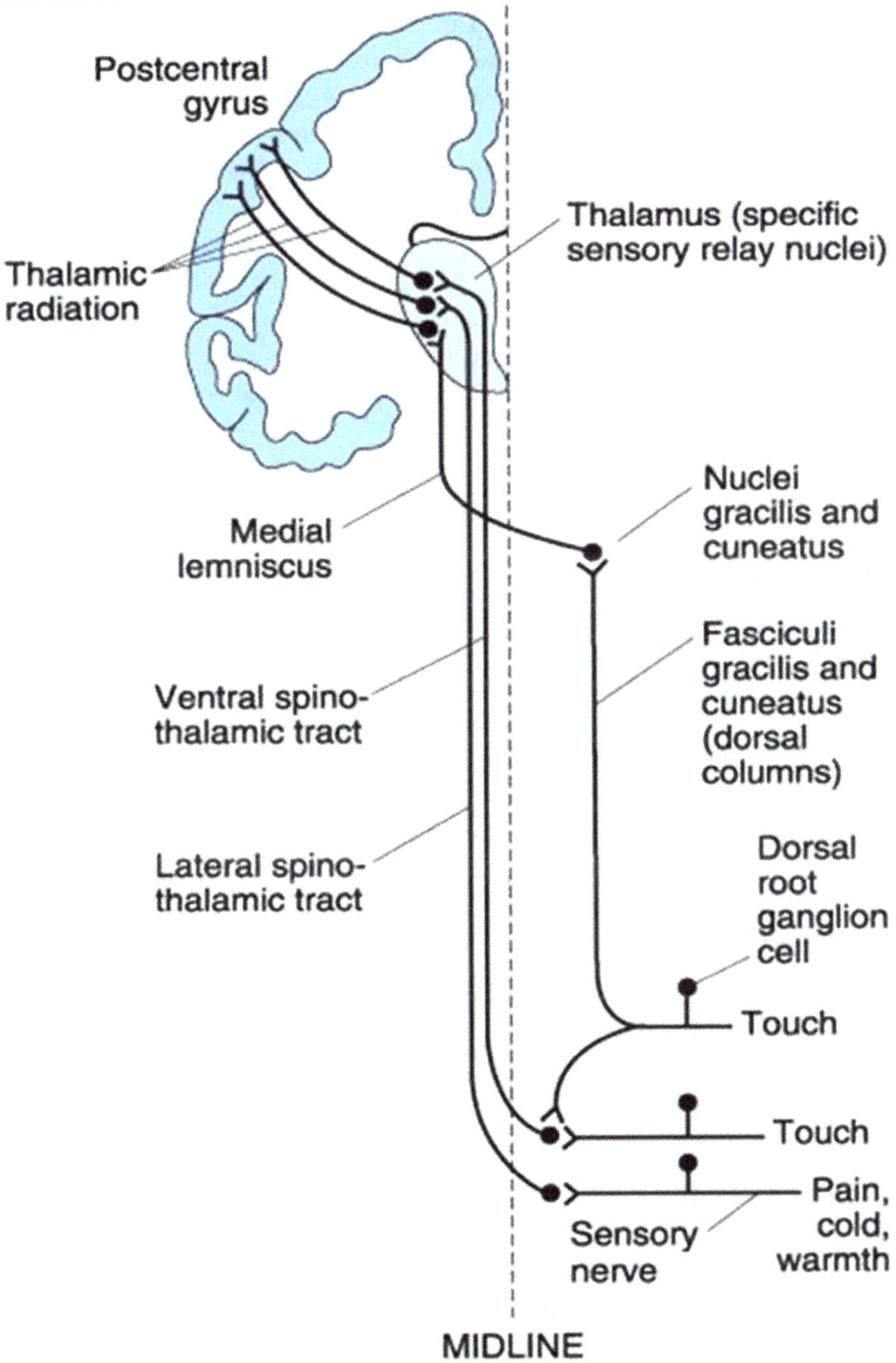

concerning discriminative senses like position sense, vibration sense, two-point discrimination, stereognosis.

- **Thalamocortical Connections:** The posterior thalamic complex receives the fibers from themedial lemniscal, trigeminal lemniscal, and spinothalamic tracts and projects mainly to two somatosensory cortical areas. The first area (S1) corresponds to the postcentral cortex or Brodmann's areas 3, 1, and 2. S1 afferents are derived primarily from the ventral posterolateral nucleus (VPL) and VPM (ventral posteromedial nucleus—the terminus of trigeminalfibers) and are distributed somatotopically, with the leg represented uppermost and the face lowermost. The second somatosensory area (S2) lies on the upper bank of the sylvian fissure, adjacent to the insula. The cortical representation of sensory information in the postcentral gyrus is shown in Figure 3 ("sensory homunculus"). A disproportionate area is devoted to localization in the fingers, lips, and face.

Clinical Implications of Sensory System Involvement

Hypesthesia is a decrease, and anesthesia an absence, of all sensation.

Hypalgesia is a decrease, and analgesia (or analgesthesia) an absence, of pain sensation.

Paresthesia is an abnormal spontaneous sensation.

Dysesthesia is an abnormal, unpleasant, or painful sensation.

Patients with stroke involving thalamus develop chronic pain on contralateral side. This is known as thalamic pain syndrome or Dejerine-Roussy syndrome.

Patients with herpes infection involving nerve root develop severe pain and dysesthesias over the corresponding dermatome. This is known as postherpetic neuralgia.

In hemisection of the spinal cord, known as Brown Sequard syndrome, there is ipsilateral weakness (corticospinal tract) and loss of joint position and vibration sense (posterior column) and contralateral loss of pain and temperature sense (spinothalamic tract) one or two levels below the lesion. Agnosia refers to a loss of recognition of an entity that cannot be attributed to a defect in the primary sensory modality.

MOTOR SYSTEM

Control of motor function – our activities like walking, lifting an object, writing etc. – is accomplished through integrated and coordinated action of motor neurons. Motor neurons can be classified into two broad categories: Upper motor neurons

Figure 2. Cross section of the spinal cord (S=sacral; L=lumbar; Th=thoracic; C=cervical segments)

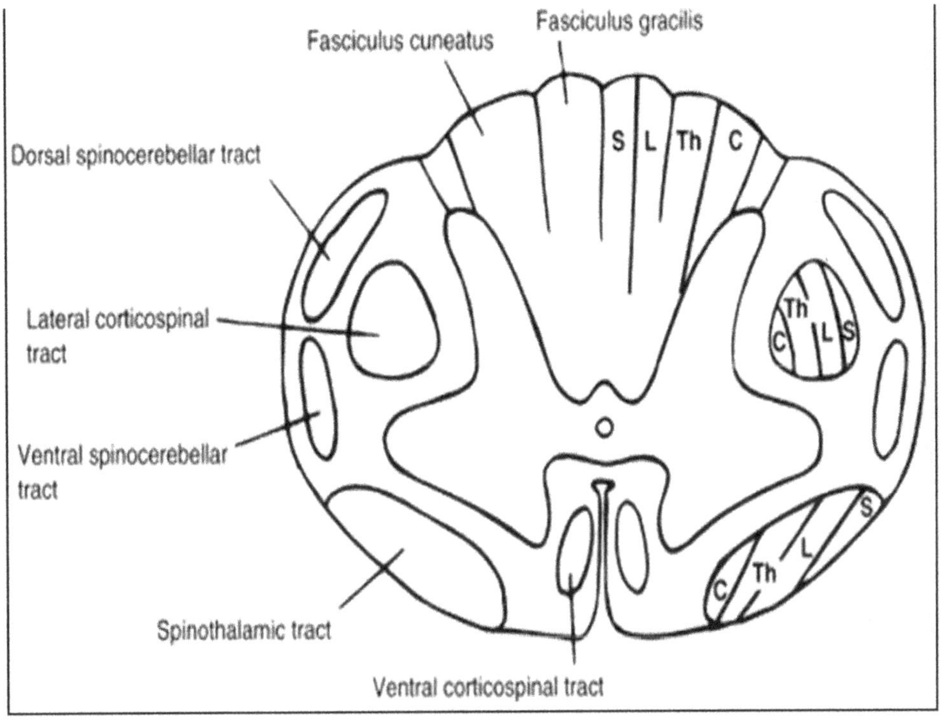

and lower motor neurons. Upper motor neurons have cell bodies in the motor cortex of brain and carry the impulses from cortex to spinal cord segment. Cell bodies of lower motor neurons are located in the anterior horn of spinal cord. Axons of lower motor neuron end in neuromuscular junction and excitation of muscle fibres leads contraction of muscle.

Control of motor function is through decending pathways in the spinal cord: the direct (pyramidal or cortico spinal) motor pathway and the indirect (extrapyramidal) system. First order neurons coming from motor cortex to anterior horn in the spinal cord are called as upper motor neurons. Second order neurons from anterior horn to muscles are called as lower motor neurons. Corticospinal tract is the principal efferent system through which voluntary movements are initiated and performed.

Motor fibres arise from primary motor cortex situated at area 4 of the precentral gyrus. This area contains giant pyramidal neurons, also known as Betz cells. Diferent parts of the body are represented in motor cortex disproportionately, with parts of body involved in skilled movements like the tongue, face, and digits having larger representation. This is depicted by motor homunculus as in Figure 4.

The term pyramidal tract arose because these fibers constitute most of the medullary pyramids. It was once thought that the pyramidal tract consisted mainly of the axons of the Betz cells in the primary motor cortex. However, only 20% to 30% arise from primary motor cortex, and only 3% arise from the Betz cells. Other fibers arise from the premotor cortex (area 6), the supplementary motor area and adjacent regions, and the postcentral gyrus (areas 3, 1, and 2) shown in figure 5. The corticospinal tract is important in controlling discrete, fine motor responses, especially fine voluntary movement of individual digits. Axons from the motor neurons of the precentral gyrus descend through the corona radiata and the posterior limb of the internal capsule shown in Figure 6.

At the midbrain level, pyramidal fibers located in the middle three-fifths of the cerebral peduncle. The majority of corticobulbar fibers cross to opposite side before synapsing with the specific cranial nerve nuclei. The descending corticospinal fibers then pass through the basilar portion of the pons and enter the medulla. In the caudal medulla, the corticospinal tracts group together as twin columns of the medullary pyramids at the base of the medulla. Approximately 85% to 90% of the fibers cross in the decussation of the pyramids, with those supplying the upper extremity decussating more rostrally than those for the lower extremity.

The fibers that decussate descend in the lateral funiculus of the spinal cord as the lateral corticospinal tract. About 50% of the fibers of the lateral corticospinal tract terminate in the cervical region, 20% in the thoracic area, and 30% in the lumbosacral portion of the cord. The uncrossed fibers decend in the ipsilateral anterior funiculus as anterior corticospinal tract. This contains about 10% to 15% of the corticospinal fibers. These fibers cross in the anterior white commissure at the segmental spinal cord level before terminating; they primarily supply axial muscles.

The axons of the corticobulbar and corticospinal tracts terminate on motor nuclei of the cranial nerves and the anterior horn cells of the spinal cord.

Clinical manifestations of disease of the corticospinal tract: The corticospinal pathways may be involved in diverse disease processes, including vascular disease, neoplasm, degeneration, trauma etc. The essential feature of corticospinal involvement is loss of skilled voluntary movements, accompanied by increased tone in the involved muscles.

A pyramidal tract lesion causes weakness in a characteristic distribution. Distal muscles, especially hand muscles, receive more pyramidal innervation than proximal muscles and are preferentially affected. In the upper limb, weakness preferentially involves the wrist, finger and elbow extensors, supinators, and external rotators and abductors of the shoulder; there is relative sparing of the flexor, pronator, and internal rotation muscles. In the lower limb, weakness is most marked in the foot and toe dorsiflexors, knee flexors, and flexors and internal rotators of the hip, with relative sparing of the extensors, external rotators, and plantar flexors.The spasticity,

Figure 3. Sensory homunculus

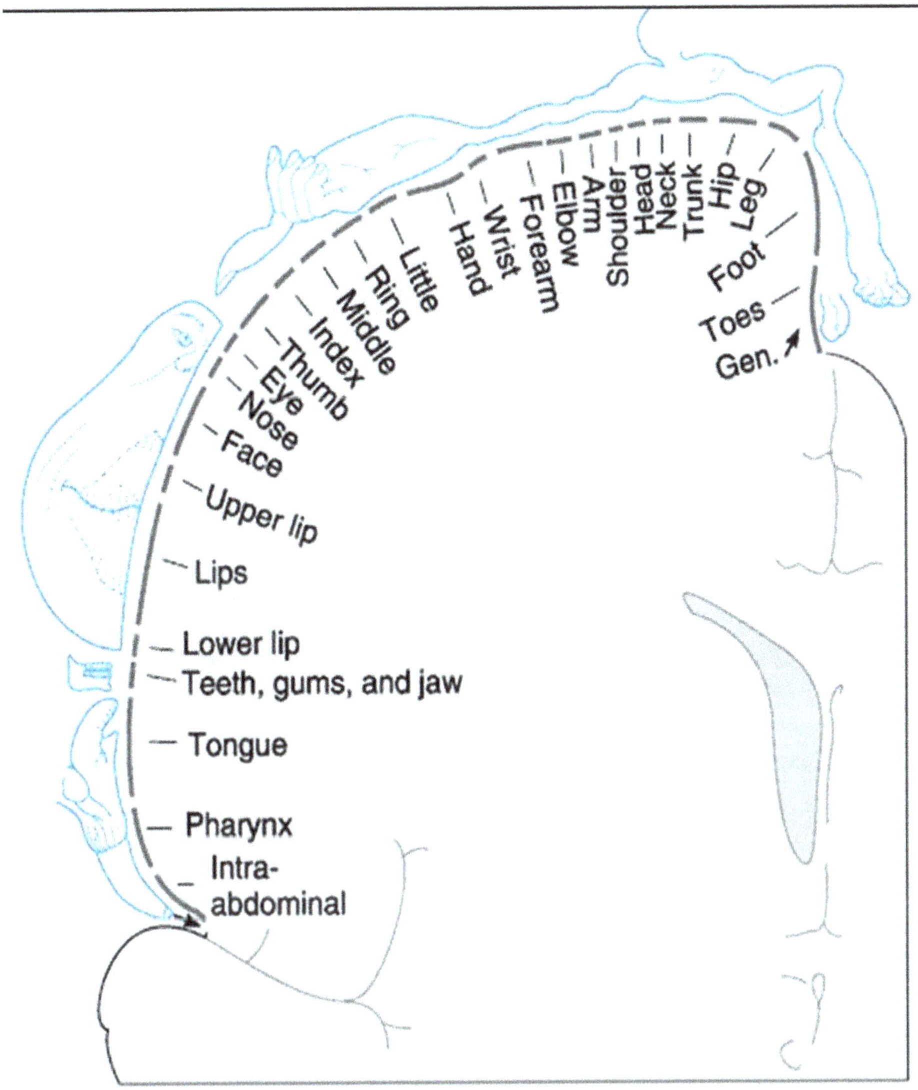

or increase in tone, is most marked in the flexor and pronator muscles of the upper limb and the extensors of the lower limb.

Hemiplegia is weakness of one half of the body. This is usually produced by upper motor neuron lesion anywhere between motor cortex to brainstem. Brain stem involvement causes crossed hemiplegia i.e. cranial nerve involvement on the same side and hemiplegia on opposite side of the lesion.

Figure 4. Motor homunculus

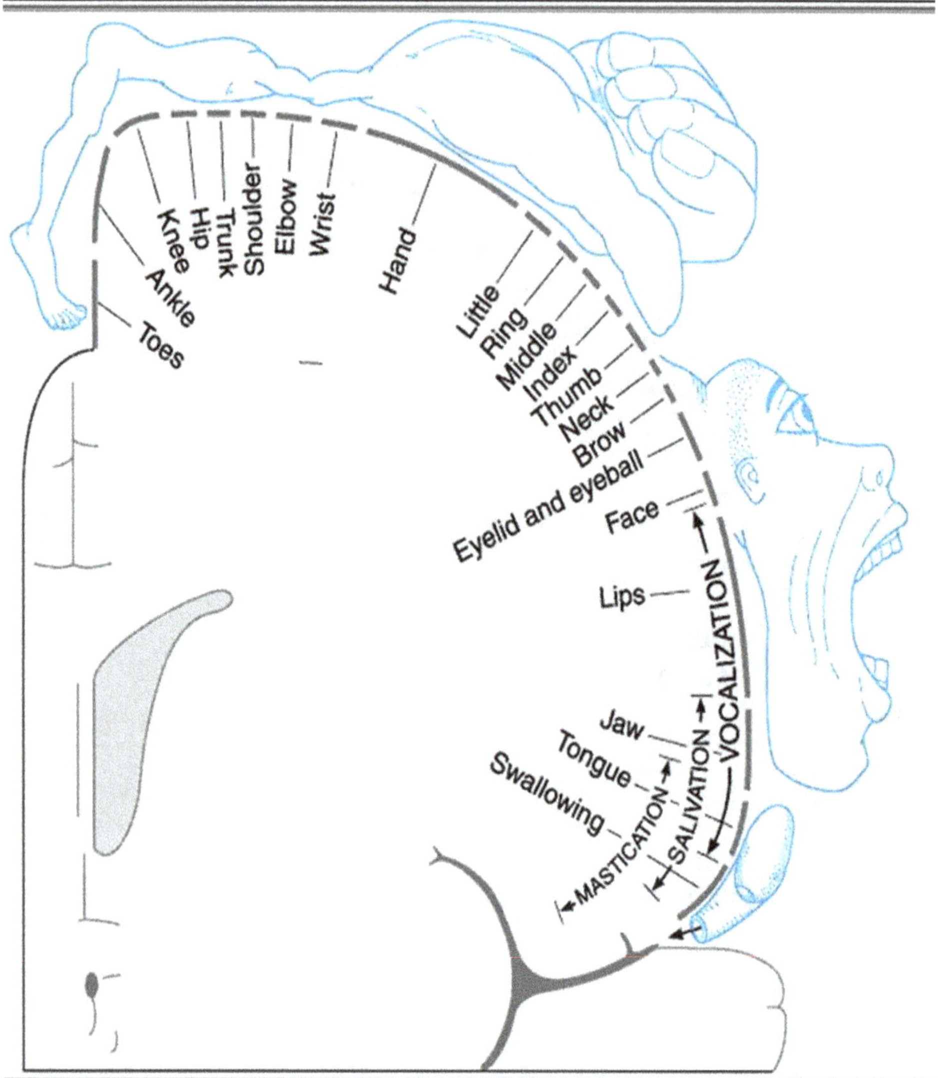

A spinal cord transection at cervical level causes quadriplegia i.e. weakness of all four limbs; whereas transaction at thoracic cord level causes paraplegia or weakness of both lower limbs accompanied by sensory loss below the level and bowel, bladder involvement. Typical example is transverse myelitis.

Figure 5. Cerebral cortex showing motor and sensory areas

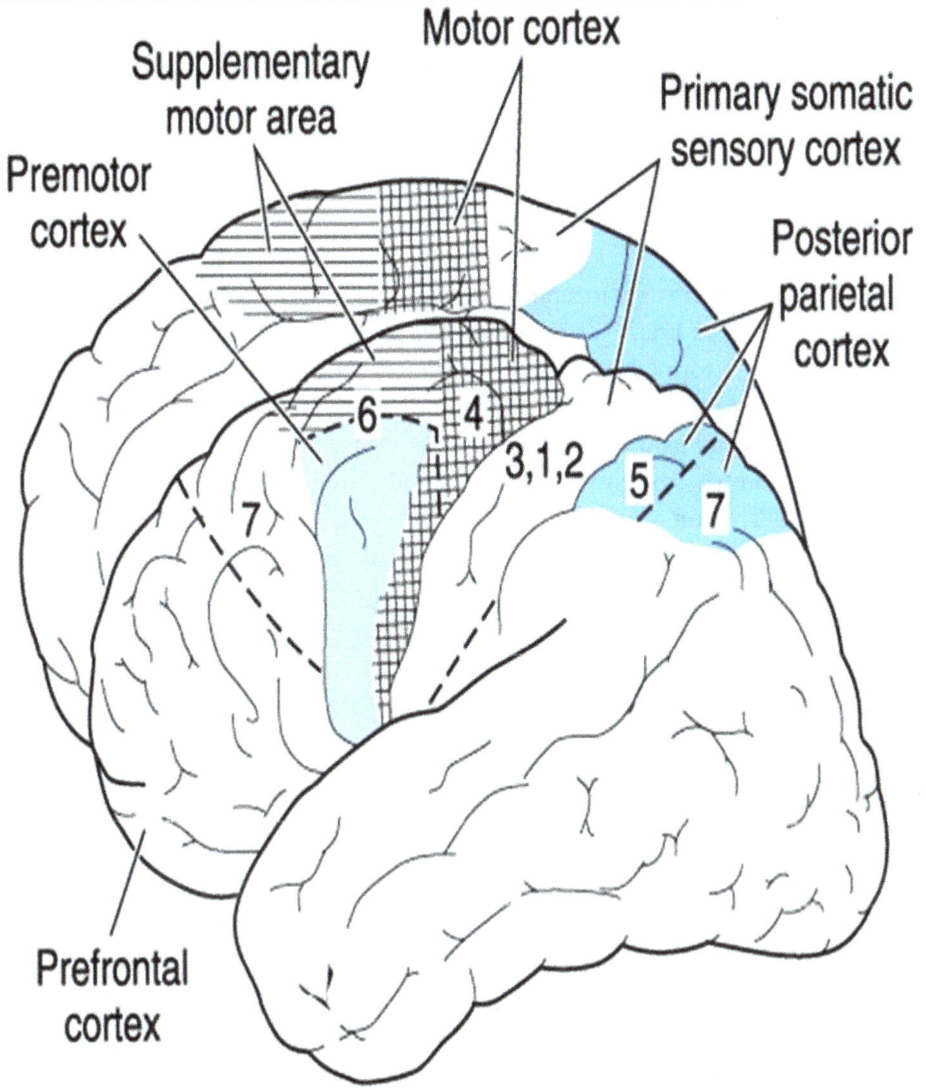

REGULATION OF INTERNAL ENVIRONMENT

The hypothalamus and limbic systems are intimately concerned with emotional expression and with the genesis of emotions. Emotions have both mental and physical components. They involve *cognition*, an awareness of the sensation and usually its cause, *affect*, the feeling itself; *conation*, the urge to take action; and *physical changes* such as hypertension, tachycardia, and sweating.

The human internal environment is regulated in large measure by the integrated activity of the autonomic nervous system and endocrine glands. Their visceral and homeostatic functions, essential to life and survival, are involuntary. The autonomic nervous system is divided into two parts: the craniosacral or parasympathetic system and the thoracolumbar or sympathetic system.

LIMBIC SYSTEM

"Limbus" is the Latin word meaning "border" or "margin". The term is used to describe the areas of brain surrounding corpus callosum and underlying upper brainstem. This is formed primarily by cingulated and parahippocampal gyri. Later term limbic lobe was extended to include underlying hippocampal formation, subcallosal gyrus and paraolfactory area.

The terms limbic system is used to include even wider structures such as amygdala, septal region, preoptic area, hypothalamus, anterior thalamus, habenula, and central midbrain tegmentum. The structures of the limbic system are involved in emotion and its expression; The major structures comprised by the limbic system are illustrated in Figure7 & Figure 8.

Clinical Manifestations of Diseases Involving Limbic System

Limbic system involvement causes following emotional abnormalities:

1. **Emotional Lability:** This refers to excessive emotional response to the event – like crying in public on meeting an old friend. This is a sign of organic brain disease and occurs in dementias like Alzheimer's disease.
2. **Pseudobulbar Affective Disorder (Emotional Incontinence):** This refers to pathologic uncontrollable laughing and crying to slightest provocation. This is seen in diseases involving corticobulbar tracts like bilateral strokes in cerebral hemispheres or pons, amyotropic lateral sclerosis, multiple sclerosis etc.
3. **Delusions and Hallucinations:** Delusions and hallucinations are disorders of thought and perception usually occur in delirium due to metabolic or toxic encephalopathy.
4. **Aggressiveness, Anger, Rage:** These emotional responses originate from temporal lobes, particularlyamygdala. Aggressive behaviour can occur as part of temporal lobe seizure. Cranial trauma can cause organic personality disorder with violent behaviour. Tumors involving temporal lobe like gliomas may present with bursts of rage. Aggressive behaviour can be seen in toxic-metabolic encephalopathies like alcohol intoxication, hypoglycaemia etc.

Figure 6. Motor pathways

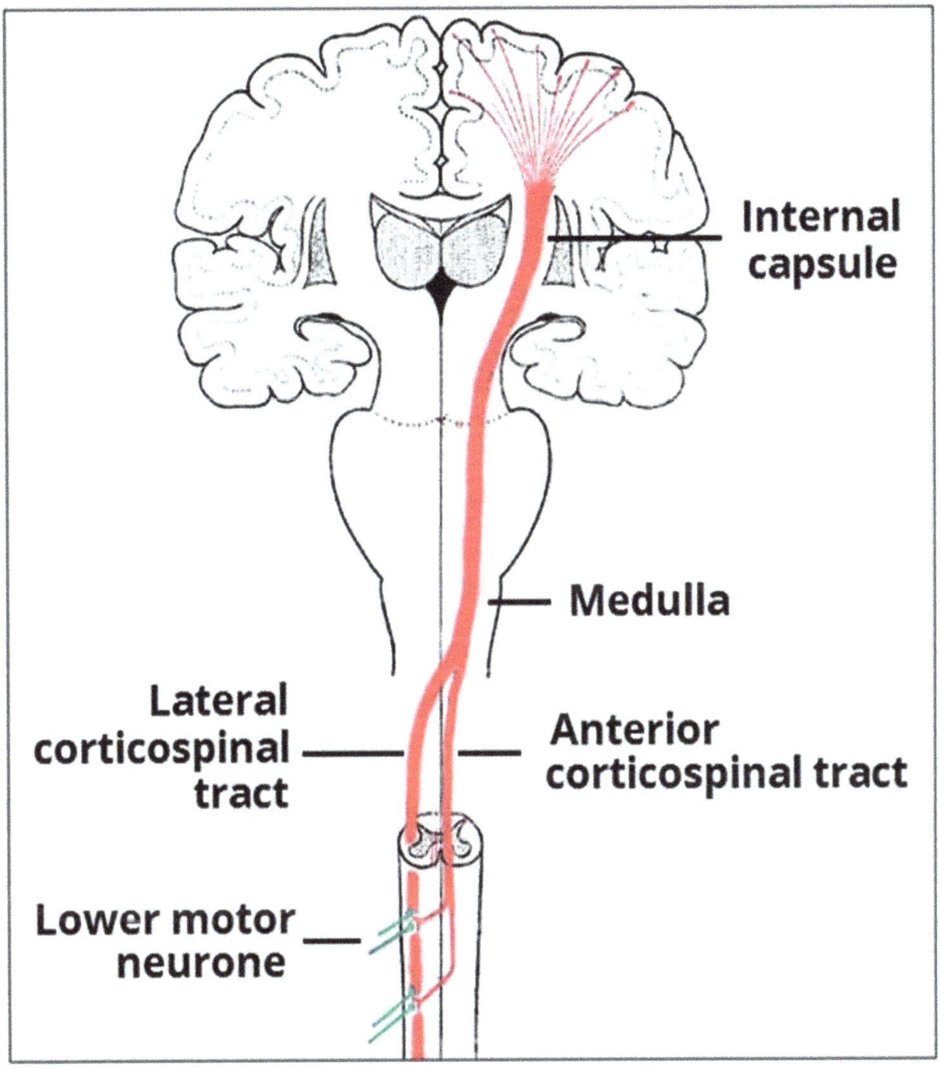

5. **Apathy:** Abulia (lack of impulse), placidity, apathy are states of reduced psychomotor activity in which the person thinks slowly, responds slowly and even the number of words spoken and movements made per unit of time are slow. Unlike depression, mood is neutral. These states are observed in diseases affecting the network involving basifrontal region like frontotemporal dementia, hydrocephalus, infarcts involving anterior cerebral artery, bleeding from ruptured anterior communicating aneurysm and frontal lobe tumors.

Figure 7. Limbic system

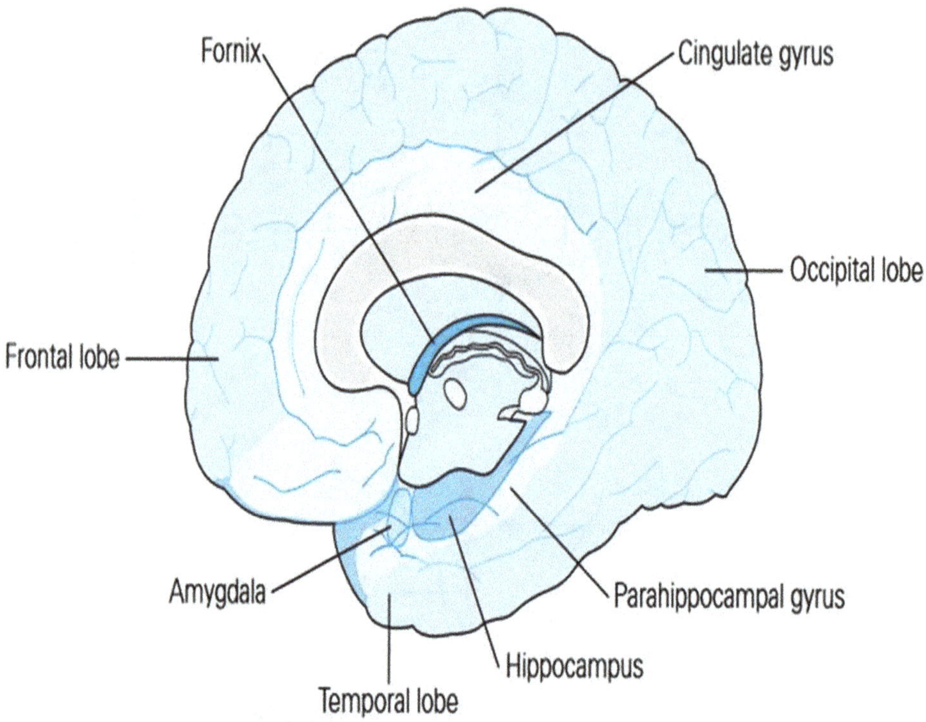

6. **Altered Sexuality:** Hypersexuality or hyposexuality can occur in diseases involving temporal lobe. Kluver Busy syndrome results from bilateral lesions of medial temporal lobes including amygdala, and is characterized by hypersexuality, hyperorality, compulsive eating and visual agnosia.
7. **Acute Fear, Anxiety, Euphoria:** These states can occur as ictal phenomenon in temporal lobe seizures.

AUTONOMIC NERVOUS SYSTEM

Human internal environment is regulated by integrated activity of autonomic nervous system. Impulses initiated in receptors located in internal organs are relayed via afferent autonomic pathways to the central nervous system, integrated within it at various levels, and then transmitted via efferent pathways to internal organs.

- **Anatomically, the Autonomic Outflow is Divided into two Components**: the *sympathetic*and *parasympathetic divisions* of the autonomic nervous system. In the gastrointestinal tract, these both communicate with the *enteric* *nervous system*, which is sometimes called a third division of the autonomic nervous system.

- **The Sympathetic and Parasympathetic Divisions are Characterized by a Two-Neuron Chain:** First order neuron or preganglionic neuron within the central nervous system that terminates in a ganglion outside the central nervous system, and a second order neuron or post-ganglionic neuron that carries impulses from ganglion to the internal organs.
- **The Parasympathetic Division is Composed of Craniosacral Outflow:** Cranial portion consisting of fibers of cranial nerves III, VII, IX, X, and XI and sacral portion consisting of fibers arising in the S2-S4 segments of the spinal cord. The parasympathetic nerves have long preganglionic fibers that end in peripheral ganglia near or in the viscera they supply, and

short postganglionic fibers that arise in proximity to or within the viscus innervated. Cranial portion of the parasympathetic division consists of the Edinger-Westphal nucleus, the superior and inferior salivatory nuclei, the dorsal motor nucleus of the vagus, and neurons in the vicinity of the nucleus ambiguus. The sacral parasympathetic fibers arise from cells in the intermediolateral cell column at the S2-S4 levels of the sacral spinal cord, travel through the sacral nerves, and are collected into the pelvic splanchnic nerves (nervi erigentes), which proceed to the pelvic plexuses and their branches. Most preganglionic fibers continue to small ganglia in or near the viscera, from where postganglionic fibers supply the bladder, descending colon, rectum, anus, and genitalia.

The sympathetic division is composed of preganglionic fibers that arise from cells in the intermediolateral columns from the T1 to the L3 segments of the spinal cord. The fibers exit through the ventral roots of the corresponding segmental nerves and terminate in the paravertebral ganglionic chain. Postganglionic fibers go to the viscera. The sympathetic preganglionic fibers are typically short and terminate on ganglia some distance from the viscera they supply, with long postganglionic fibers that travel from the ganglia to the viscera.

The T1-T3 segments innervate the head and neck, the T3-T11 segments innervate the upper extremities and viscera in the thorax and abdomen, and the T12-L2 segments innervate the lower extremities and pelvic viscera.

The sympathetic nervous system prepares the organism for fight-or-flight response. It acts whenever rapid adjustment to the environment is required. It accelerates the heart, dilates the coronary vessels, increases the arterial blood pressure, dilates the bronchi, releases glucose, and inhibits GI activity. It is an emergency protective mechanism that is called into action under emotional stress and causes the individual to react strongly to external stimuli. The parasympathetic nervous system conserves energy. It controls anabolic, excretory, and reproductive functions, and conserves and restores bodily resources. The internal organs receive a dual autonomic supply, both

Figure 8. Limbic connections

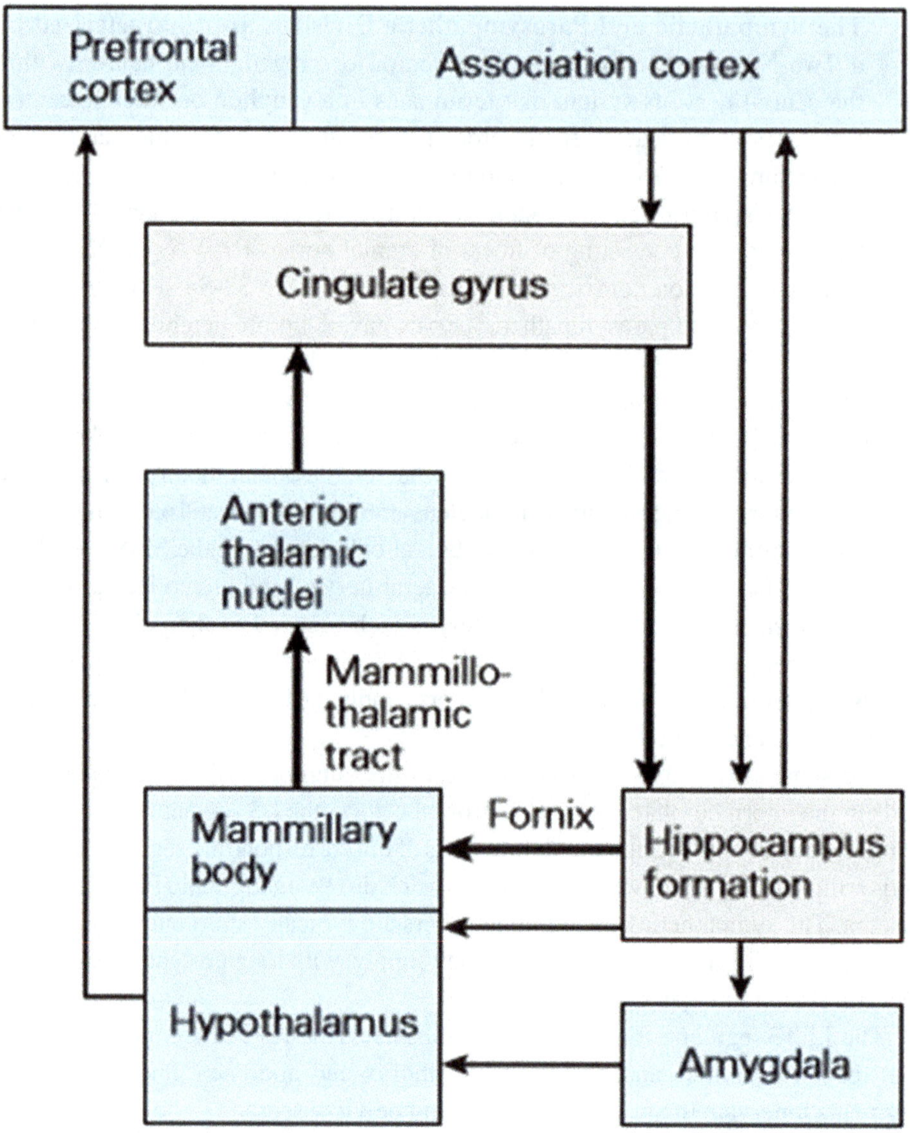

sympathetic and parasympathetic. In general these two divisions are antagonistic and reciprocal in their functions.

Clinical Disorders of the Autonomic Nervous System

- **Acute Autonomic Failure (Pan Dysautonomia):** In this condition, both sympathetic and parasympathetic post ganglionic fibres are affected. It ca be idiopathic, post infectious or paraneoplastic. Symptoms of anhydrosis, orthostatic hypotension, loss of lacrimation and salivation, impotence, impaired bowel and bladder function develop over days to weeks. Autonomic dysfunction as part of peripheral neuropathies: Many acute and chronic peripheral neuropathies like diabetic, amyloid, alcohol-nutrtional, Guillain-Barre syndrome, toxic, porphyric neuropathies are associated with prominent autonomic dysfunction including postural hypotension. Diabetic neuropathy presents as impotence, constipation or diarrhea, hypotonia of the bladder, gastroparesis, and orthostatic hypotension in addition to signs of a sensory polyneuropathy, like distal sensory loss to pain-temperature and vibration, loss of ankle jerks. Amyloid polyneuropathy also has prominent autonomic dysfunction in addition to loss of pain and temperature sense.
- **Horner Syndrome:** Horner syndrome or Bernard-Horner syndrome refers to combination of miosis, drooping of the eyelid, and loss of sweating over one side of the face due to damage to postganglionic sympathetic fibers along the internal carotid arteries or a lesion of the superior cervical ganglion.
- **Multiple System Atrophy:** Orthostatic hypotension can occur as part of neurodegenerative disorder known as multiple system atrophy, along with anhidrosis, impotence, and atonicity of the bladder.
- **Bladder Dysfunction:** Functioning of bladder is controlled by parasympathetic, sympathetic and motor innervation. Parasympathetic supply is from sacral segments S2,S3,S4. Sympathetic supply is from T10,T11,T12 segments. Motor innervations is through pudendal nerve derived from nucleus of Onuf formed from sacral segments S2,S3,S4.

In complete destruction of the spinal cord below T12, both voluntary and reflex activities of bladder are lost. There will be no awareness of bladder fullness, voluntary initiation of micturition is not possible. There will be urinary retention with overflow incontinence.

Upper spinal cord lesions above T12 will result in reflex neurogenic or spastic bladder. Bladder capacity is reduced and there will be urgency and urge incontinence.

THE CENTRAL REGULATION OF AUTONOMIC FUNCTION

Neuroscientists Cannon, Bard, and others established the fact that the hypothalamus contains the suprasegmental integrations of the autonomic nervous system, both the sympathetic and parasympathetic.Efferent pathways from the hypothalamus to centers subserving parasympathetic and sympathetic reflexes were found later. Cannon's emergency theory of sympathoadrenal action was based upon these gmental reflex, involving the sympathetic innervation of the adrenal gland. Later, the hypothalamus was also found to contain neurosecretory cells, which control the secretion of the pituitary hormones. Special sensory receptors for there gulation of hunger, thirst, body temperature, and levels of circulating electrolytes are also located in hypothalamus. Gradually the idea of a hypothalamic-pituitary-autonomic system emerged that is essential for the basic homeostatic and emergency reactions. Apart from hypothalamus, other higher centers in the cerebral cortex, especially the amygdala, basal forebrain, ventral striatum, brainstem, and spinal cord also regulate and influence the function of peripheral components of autonomic nervous system. The hypothalamus plays three roles in the actions of the nervous system. The first, as the "head ganglion" of the autonomic nervous system; the second, as the circadian and seasonal clock for behavioral and sleep-wake functions; and the third, as the neural center of the endocrine system.

The Supranuclear Regulatory Apparatus of the Hypothalamus Includes Three Main Cerebral Structures: the frontal lobe cortex, the insular cortex, and the amygdaloid and adjacent nuclei. These areas function as the highest levels of autonomic integration.

Other important centers involved in autonomic control include the periaqueductal gray matter (PAG) in the midbrain, other brainstem nuclei, the cerebral cortex, and the amygdala. The periaqueductal gray matter is important in the micturition reflex, pain mechanisms and the fight-or-flight response. The nucleus of the solitary tract (NST) in the medulla is involved in cardiopulmonary and gastrointestinal function. The medullary cardiorespiratory centers control blood pressure and respiration and mediate cardiorespiratory reflexes.

The primary cortical areas involved in autonomic function include the cortex of the insula, the medial prefrontal cortex, the cingulate gyrus, and the nucleus of the amygdala. Sensory input from the viscera project to the insula. The amygdala communicates with the hypothalamus, periaqueductal gray matter and the brainstem autonomic nuclei.

NEUROENDOCRINE SYSTEM

- **Hypothalamus:** The hypothalamus plays three roles in the actions of the nervous system. The first, as the "head ganglion" of the autonomic nervous system, as described earlier; the second, as the circadian clock for behavioral and sleep-wake functions; and the third, as the neural center of the endocrine system. The hypothalamus lies on each side of the third ventricle and is continuous across its floor. It is bounded posteriorly by the mammillary bodies, anteriorly by the optic chiasm, superiorly by the hypothalamic sulci, laterally by the optic tracts, and inferiorly by the hypophysis (pituitary). It comprises three main nuclear groups: (1) the anterior group, which includes the preoptic, supraoptic, and paraventricular nuclei; (2) the middle group, which includes the tuberal, arcuate, ventromedial, and dorsomedial nuclei; and (3) the posterior group, comprising the mammillary and posterior hypothalamic nuclei.
- **The Hypothalamic-Pituitary Axis:** Hypothalamus is connected to pituitary gland by pituitary stalk. Pituitary gland is divided functionally in to anterior adenohypophysis which secretes most of the pituitary hormones and posterior neurohypophysis, which contains axons of supraoptic and paraventricular neurons of hypothalamus.
- **Hypothalamic Releasing Hormones:** Thyrotropin releasing hormone (TRH): TRH stimulates the release of thyroid stimulating hormone (TSH) from the pituitary gland; TSH, in turn, effects the release of T4 (thyroxine) and T3 (triiodothyronine).
- **Growth Hormone Releasing Hormone (GHRH):** This hormone and somatostatin (also known as GH-release inhibiting hormone, or SRIH) are secreted by tuberoinfundibular neurons and released into the hypophysial-portal circulation, by which they are carried to specific GH-secreting cells of the anterior pituitary gland (somatotropes).
- **Corticotropin Releasing Hormone (CRH):** This hormone, stimulates release of adreno cortico tropic hormone (ACTH)from the pituitary. ACTH stimulates the synthesis and release of the hormones of the adrenal cortex, mainly glucocorticoids.
- **Gonadotropin Releasing Hormone (GnRH***):* It effects the release of the two gonadotropic hormones luteinizing hormone (LH) and follicle-stimulating hormone (FSH). Puberty, menstruation, ovulation, lactation, and menopause are all related to the effects of GnRH, FSH, and LH on the ovaries, uterus, breasts, and testes.
- **Prolactin Releasing Hormone (PRH):** This hormone is released by neurons in the region of the arcuate nucleus and activates lactotropic cells of the

anterior pituitary. Sensory stimuli from the nipples, via pathways in the spinal cord and brainstem, activate hypothalamic pituitary axis leading to milk production. The release of prolactin is inhibited by dopaminergic innervation of pituitary lactotrophic cells. Tumors that compress the pituitary stalk interrupt the venous portal transport of dopamine from the hypothalamus to pituitary leading to galactorrhea and reproductive disorders.

- **The Neurohypophysis**: Vasopressin and oxytocin are secreted by cells of the supraoptic and paraventricular nuclei and are transported, via their axons, through the stalk of the pituitary to its posterior lobe, where these substances are stored. Together, these elements constitute the neurohypophysis (posterior pituitary). Vasopressin, acting on the kidney tubules, serves as the antidiuretic hormone (ADH) and, complemented by thirst mechanisms, maintains the osmolality of the blood. Oxytocin initiates uterine contraction and has milk-ejecting effects

Thus, hypothalamus plays important role in the integration of the endocrine and autonomic nervous systems at both the peripheral and central levels.

REFERENCES

Brazis, P. W., Masdeu, J. C., & Biller, J. (2007). *Localization in clinical neurology*. Philadelphia: Lippincott Williams & Wilkins.

Campbell, W. W., DeJong, R. N., & Haerer, A. F. (2005). *DeJong's the neurologic examination: Incorporating the fundamentals of neuroanatomy and neurophysiology*. Philadelphia, PA: Lippincott Williams & Wilkins.

Daroff, R. B., & Bradley, W. G. (2012). *Bradley's neurology in clinical practice*. Philadelphia, PA: Elsevier/Saunders.

Ropper, A. H., Adams, R. D., Victor, M., & Brown, R. H. (2005). Adams and Victor's principles of neurology (8th ed.). New York: McGraw-Hill Medical Pub. Division.

KEY TERMS AND DEFINITIONS

Dorsal Root Ganglia: Cluster of cell bodies of sensory neurons in dorsal root of spinal nerves.

Hypothalamus: Part of the brain below thalamus containing small nuclei with various functions, forms link between nervous system and endocrine system via pituitary gland.

Limbic System: Set of brain structures involved in behaviour, olfaction, and long-term memory.

Lower Motor Neuron: Neurons that originate anterior horn of spinal cord or cranial nerve nuclei in brainstem and innervate skeletal muscles.

Parasympathetic Nervous System: One of the two main divisions of autonomic nervous system which is responsible for stimulation of "rest and digest" or "feed and breed."

Sympathetic Nervous System: One of the two main divisions of autonomic nervous system which stimulates "fight or flight" response.

Upper Motor Neuron: Neurons that originate in the cerebral motor cortex and carry impulses to inter neurons or lower motor neurons.

Chapter 6
Neurotransmitter and Behaviour

Manoj Kumar Bajaj
https://orcid.org/0000-0002-9999-7888
Government Medical College and Hospital, Chandigarh, India

ABSTRACT

Neurotransmitters play a major role in everyday life and functioning. Everything known about human behaviour suggests it is regulated entirely by the human brain. Brain cells (neurons) communicate with one another (synaptic transmission) and with other cells in the body through small molecules called neurotransmitters (NT). NT are released by neurons and picked up by targeted cells through NT receptors (NTR). Increase or decrease in the production of any of these molecules due to any reason can produce profound effects on behaviour. Knowledge of the pathways involved in NT function has allowed development of drugs that modulate these pathways up or down. Scientists do not yet know exactly how many neurotransmitters exist, but more than 200 chemical messengers have been identified.

CHAPTER OUTLINE

- Introduction
- Major neurotransmitters
- Disorders associated with NT systems
- Serotonin
- Glutamate
- Acetylcholine
- Neurotransmitter basis of learning

DOI: 10.4018/978-1-7998-2860-0.ch006

- Neurotransmitters adverse impacts
- Neurotransmitter Dysfunction in Behavioural Disorder
- Inhibitory Neurotransmitters
- Neurotransmitters influenced by the various drugs
- Conclusion

This chapter highlight the role of relationship of neurotransmitters with behaviour. It includes specifically neurotransmitters classification according to its chemical structure and mechanism of actions, its types etc. Neurotransmitters mechanism of action in various mental disorders is discussed in details to help reader develop the understanding while dealing such conditions in their practice. Students of clinical psychology discipline will have an organised and important understanding and its use in various conditions, its therapeutic role in treating various anxiety, emotional and behavioural disorders.

INTRODUCTION

Neurotransmitters play a major role in everyday life and functioning. Everything known about human behaviour suggests it is regulated entirely by the human brain. Brain cells (Neurons) communicate with one another (synaptic transmission), and with other cells in the body, through small molecules called neurotransmitters (NT). NT is released by neurons, and is picked up by targeted cells through NT receptors (NTR). Increase or decrease in the production of any of these molecules due to any reasons can produce profound effects on behaviour. Knowledge of the pathways involved in NT function has allowed development of drugs that modulate these pathways up or down. Scientists do not yet know exactly how many neurotransmitters exist, but more than 200 chemical messengers have been identified.

For the simple and better understanding of these functions is presented in the following sections.

There are many different ways to classify neurotransmitters. Dividing them into amino acids, peptides (Snyder & Innis, 1979), and monoamines is sufficient for some classification purposes.

MAJOR NEUROTRANSMITTERS

- **Amino Acids:** glutamate, aspartate, D-serine, γ-amino butyric acid (GABA), glycine
- **Monoamines:** dopamine (DA), norepinephrine (noradrenaline;NE,NA), epinephrine (adrenaline), histamine, serotonin (SER, 5-HT)
- **Peptides:** oxytocin, somatostatin, substance P, cocaine and amphetamine regulated transcript, opioid peptides.
- **Others:** acetylcholine (Ach) etc.

Neurons expressing certain types of neurotransmitters sometimes form distinct systems, where activation of the system affects large volumes of the brain, called volume transmission. Major neurotransmitter systems include the noradrenaline (norepinephrine) system, the dopamine system, the serotonin system, and the cholinergic system, among others. These specific neurotransmitter systems have specific roles in different cognitive, behavioural, emotion and autonomic functions are shown in table1.

DISORDERS ASSOCIATED WITH NT SYSTEMS

Diseases and disorders may also affect specific neurotransmitter systems. The following are disorders involved in an increase, decrease, or imbalance of certain neurotransmitters.

DOPAMINE

Dopamine has a number of important functions in the brain; this includes regulation of motor behavior, pleasures related to motivation and also emotional arousal. It plays a critical role in the reward system; Parkinson's disease has been linked to low levels of dopamine and schizophrenia has been linked to high levels of dopamine (Schacter, *Gilbert & Wegner,* 2009). Dopamine decreased production (mainly in the substantia nigra) can result in Parkinson's disease, a disorder that affects a person's ability to move as they want to, resulting in stiffness, tremors or shaking, and other symptoms. Some studies suggest that having too little or too much dopamine or problems using dopamine in the thinking and feeling regions of the brain may play a role in disorders like schizophrenia or attention deficit hyperactivity disorder (ADHD). Dopamine is also involved in addiction and drug use, as most recreational drugs cause an influx

Table 1.

Noradrenaline system	Dopamine system
Anxiety Arousal (wakefulness) Circadian rhythm Cognitive control and working memory (co-regulated by dopamine) Feeding and energy homeostasis Medullary control of respiration Negative emotional memory Nociception (perception of pain) Reward (minor role)	Arousal (wakefulness) Aversion Cognitive control and working memory (co-regulated by norepinephrine) Emotion and mood Motivation (motivational salience) Motor function and control Positive reinforcement Reward (primary mediator) Sexual arousal, orgasm, and refractory period (via neuroendocrine regulation)
Serotonin system	**Acetylcholine system**
Arousal (wakefulness) Body temperature regulation Emotion and mood, potentially including aggression Feeding and energy homeostasis Reward Sensory perception	Arousal (wakcfulncss) Emotion and mood Learning Motor function Motivation (motivational salience) Short-term memory Reward

of dopamine in the brain especially opioid and methamphetamines that produces a pleasurable feeling, which is why users constantly crave drugs.

SEROTONIN

Serotonin is a monoamine neurotransmitter. Most is produced by and found in the intestine (approximately 90%), and the remainder in central nervous system neurons. It functions to regulate appetite, sleep, memory and learning, temperature, mood, behaviour, muscle contraction, and function of the cardiovascular system and endocrine system. It is speculated to have a role in depression, as some depressed patients are seen to have lower concentrations of metabolites of serotonin in their cerebrospinal fluid and brain tissue (Miller, 2011). Some research suggested that drugs that block the recycling, or reuptake, of serotonin seemed to help some people diagnosed with depression, it was theorized that people with depression might have lower-than-normal serotonin levels. Though widely popularized, this theory was not borne out in subsequent research (Van Praag, 1981). Therefore, selective serotonin reuptake inhibitors (SSRIs) are used to increase the amounts of serotonin in synapses. A class of drugs called SSRIs can regulate the flow of the NT serotonin throughout the brain, affecting disorders such as depression and impulsivity.

GLUTAMATE

Glutamate is used at the great majority of fast excitatory synapses in the brain and spinal cord. It is also used at most synapses that are "modifiable", i.e. capable of increasing or decreasing in strength. Modifiable synapses are thought to be the main memory-storage elements in the brain. Excessive glutamate release can overstimulate the brain and lead to excitotoxicity causing cell death resulting in seizures or strokes (Gross, 2006). Excitotoxicity has been implicated in certain chronic diseases including ischemic stroke, epilepsy, amyotrophic lateral sclerosis, Alzheimer's disease, Huntington disease, and Parkinson's disease (Yang et al. 2011). Problems with producing or using glutamate have been suggestively and tentatively linked to many mental disorders, including autism, obsessive compulsive disorder (OCD), schizophrenia, and depression (Healy, 2015). NIMH Brain Basics, 2014 reports having too much glutamate has been linked to neurological diseases such as Parkinson's disease, multiple sclerosis, Alzheimer's disease, stroke, and ALS (amyotrophic lateral sclerosis). The NT glutamate and its receptor are critical in learning and memory.

ACETYLCHOLINE

Acetylcholine was the first neurotransmitter discovered in the peripheral and central nervous systems. It activates skeletal muscles in the somatic nervous system and may either excite or inhibit internal organs in the autonomic system (Wishaw et al., 2014). It is distinguished as the transmitter at the neuromuscular junction connecting motor nerves to muscles. The paralytic arrow-poison curare acts by blocking transmission at these synapses. Acetylcholine also operates in many regions of the brain, but using different types of receptors, including nicotinic and muscarinic receptors. Acetylcholine is the substance the nervous system uses to activate skeletal muscles, a kind of striated muscle. These are the muscles used for all types of voluntary movement, in contrast to smooth muscle tissue, which is involved in a range of involuntary activities such as movement of food through the gastrointestinal tract and constriction of blood vessels. Skeletal muscles are directly controlled by motor neurons located in the spinal cord or, in a few cases, the brainstem. These motor neurons send their axons through motor nerves, from which they emerge to connect to muscle fibers at a special type of synapse called the neuromuscular junction. When a motor neuron generates an action potential, it travels rapidly along the nerve until it reaches the neuromuscular junction, where it initiates an electrochemical process that causes acetylcholine to be released into the space between the presynaptic terminal and the muscle fiber. The acetylcholine molecules then bind to nicotinic ion-channel receptors on the muscle cell membrane, causing the ion channels to open. Sodium ions then flow into the muscle cell, initiating a sequence of steps that finally produce muscle contraction. Acetylcholine has also been implicated in learning and memory in several ways. The anti-cholinergic drug, scopolamine, impairs acquisition of new information in humans (Yang et al., 2011) and animals (Kodirov et al., 2006). In animals, disruption of the supply of acetylcholine to the neocortex impairs the learning of simple discrimination tasks, comparable to the acquisition of factual information and disruption of the supply of acetylcholine to the hippocampus and adjacent cortical areas produces forgetting comparable to anterograde amnesia in humans. The disease myasthenia gravis, characterized by muscle weakness and fatigue, occurs when the body inappropriately produces antibodies against acetylcholine nicotinic receptors, and thus inhibits proper acetylcholine signal transmission. Over time, the motor end plate is destroyed. Drugs that competitively inhibit acetylcholinesterase (e.g., neostigmine, physostigmine, or primarily pyridostigmine) are effective in treating this disorder. They allow endogenously released acetylcholine more time to interact with its respective receptor before being inactivated by acetylcholinesterase in the synaptic cleft (the space between nerve and muscle).

GABA is used at the great majority of fast inhibitory synapses in virtually every part of the brain. Many sedative/tranquilizing drugs act by enhancing the effects of GABA. Correspondingly, glycine is the inhibitory transmitter in the spinal cord.

Norepinephrine which is synthesized in the central nervous system and sympathetic nerves, modulates the responses of the autonomic nervous system, the sleep patterns, focus and alertness. It is synthesized from tyrosine.

Epinephrine which is also synthesized from tyrosine is released in the adrenal glands and the brainstem. It plays a role in sleep, with one's ability to become and stay alert, and the fight-or-flight response.

Histamine works with the central nervous system (CNS), specifically the hypothalamus (tuberomammillary nucleus) and CNS mast cells.

NEUROTRANSMITTER BASIS OF LEARNING

One of the most exciting areas of research is the attempt to find out how learning and memory take place. One of the earliest researchers who attempted to explain learning and memory as a function of cellular change was the Canadian psychologist, Donald O. Hebb. He maintained that repeated firing of axons results in metabolic changes in the presynaptic and post synaptic neurons. In other words, learning produces lasting chemical changes in nerve cells. Aplysia, a marine snail with only 20,000 relatively large neurons, has been used in studies to determine the biological basis of learning. A conditioned reflex in Apylsia has been shown to cause an increase in the release of a neurotransmitter that sets up a chain of reactions, one of which also increases the secretion of serotonin from a modulating neuron. In mammals, the hippocampus, part of the forebrain, stores long term memory for weeks before transferring it to the cerebral cortex. The transmitter used for long term potentiation is the amino acid, glutamate, which binds to receptors in the postsynaptic cell. This cell allows calcium to flow in and set up the activation of other molecules known as kinases (Kuffler et al.1976). A growing body of evidence implicates the role of dopamine as one of the most important chemicals that regulate cell activity involved in working memory. Studies done with aged monkeys show that a deficiency in both dopamine and norepinephrine in the prefrontal cortex can induce a deficit in working memory. Injections of the deficient neurotransmitters restored memory function. Progress in deciphering the operation of the nervous system has helped to increase knowledge of the diverse role of the neurotransmitter (Raven et al., 1992).

NEUROTRANSMITTERS ADVERSE IMPACTS

As with many of the body's processes, things can sometimes go awry. It is perhaps not surprising that a system as vast and complex as the human nervous system would be susceptible to problems. It may cause:

- Neurons might not manufacture enough of a particular neurotransmitter
- Too much of a particular neurotransmitter may be released
- Too many neurotransmitters may be deactivated by enzymes
- Neurotransmitters may be reabsorbed too quickly

When neurotransmitters are affected by disease or drugs, there can be a number of different adverse effects on the body. Diseases such as Alzheimer's, epilepsy, and Parkinson's are associated with deficits in certain neurotransmitters. Medications are sometimes used alone, but they may also be used in conjunction with other therapeutic treatments including cognitive behaviour therapy.

NEUROTRANSMITTER DYSFUCNTION IN BEHAVIOURAL DISORDER

Abnormalities in the regulation of neurotransmitter release and/or abnormal levels of extracellular neurotransmitter concentrations have remained core components of hypotheses on the neuronal foundations of behavioral and cognitive disorders and the symptoms of neuropsychiatric and neurodegenerative disorders. Furthermore, therapeutic drugs for the treatment of these disorders have been developed and categorized largely on the basis of their effects on neurotransmitter release and resulting receptor stimulation. Several examples illustrate that the nature of the neurotransmitter dysregulation in animal models, including the direction of drug effects on neurotransmitter release, depends fundamentally on the state of activity of the neurotransmitter system of interest and on the behavioral and cognitive functions recruiting these systems. Contemporary hypothesis describing neurotransmitter abnormalities underlying neuropsychiatric disorders have continued to characterize these abnormalities in terms of steady-state deviations from normal levels of neurotransmitter release or extracellular concentrations of neurotransmitters. For example: too much dopamine in critical regions in the brain, or too little glutamate…' (Fischbach, 2006; p. 734) has been widely hypothesized to contribute to the manifestation of the psychotic and cognitive symptoms of schizophrenia (see also Laruelle et al, 2003). Likewise, increased or sensitized release of dopamine in striatal regions has been closely linked with compulsive drug seeking behavior (eg, Everitt and Robbins, 2005; Robinson and Berridge, 2003). Other examples include the abnormally low levels of cholinergic neurotransmission, considered to represent an essential variable in the manifestation of dementia (Mesulam, 2004), or abnormally high levels of cholinergic transmission, reflecting a disinhibited cortical cholinergic input system, which has been hypothesized to contribute to the impairments in the filtering of irrelevant stimuli in schizophrenia (Sarter, 1994; Sarter et al, 2005). Furthermore, the antidepressant actions of a wide range of compounds acting on opioid, galanin, or cannabinoid receptors have formed the basis for suggestions that abnormal levels of these neurotransmitters/neuromodulators contribute to the symptoms of depression. In addition, the demonstration of the behavioral and cognitive symptoms, as a result of the administration of drugs that attenuate increases in neurotransmitter release or their postsynaptic effects, such as the detrimental cognitive effects of nicotinic and muscarinic receptor antagonists, or the depression-like symptoms resulting from acute tryptophan depletion (eg, Booij et al, 2005; Ellis et al, 2006), have been interpreted as supporting the hypothesis that abnormal levels of neurotransmitter release play a major role in the development of the cognitive, behavioral, and affective symptoms of major disorders.

INHIBITORY NEUROTRANSMITTERS

Inhibitory neurotransmitters: These types of neurotransmitters have inhibitory effects on the neuron; they decrease the likelihood that the neuron will fire an action potential. Some of the major inhibitory neurotransmitters include serotonin and *gamma-aminobutyric acid (GABA)*. GABA is synthesised from ***glutamate*** and is an inhibitory neurotransmitter within the CNS. ***Inhibitory*** neurotransmitters decrease electrical excitability on the post-synaptic side to prevent propagation of an action potential. It binds to two different receptors:

- GABA A receptors – ***ionotropic*** receptors permeable to chloride and bicarbonate ions
- GABA B receptors – ***metabotropic*** G protein coupled receptors

GABA has both rapid inhibitory effects when binding to post-synaptic receptors and slower inhibition via ***neuromodulation*** at pre-synaptic receptors. It is involved in many different processes in the brain, such as regulating neuronal activity; anxiety and sleep.

GLYCINE

Glycine is an amino acid which is used at the majority of inhibitory synapses in the spinal cord and brainstem. It binds to ***ionotropic*** receptors which are permeable to chloride and bicarbonate ions. As an inhibitory neurotransmitter glycine is important in many motor and sensory functions, such as reciprocal inhibition of antagonistic muscles in spinal reflexes. Glycine also has an excitatory role within the CNS as it is a ***co-agonist*** at glutamatergic NMDA receptors.

GABA and Anxiety Disorders

The brain circuits in the amygdala are thought to comprise inhibitory networks of γ-aminobutyric acid-ergic (GABAergic) interneurons and this neurotransmitter thus plays a key role in the modulation of anxiety responses both in the normal and pathological state. The presence of allosteric sites on the $GABA_A$ receptor allows the level of inhibition of neurons in the amygdala to be regulated with exquisite precision, and these sites are the molecular targets of the principal classes of anxiolytic drugs. Changes in the levels of endogenous modulators of these allosteric sites as well as changes in the subunit composition of the $GABA_A$ receptor may represent mechanisms whereby the level of neuronal inhibition is downregulated in pathological anxiety states. Neurosteroids are synthesized in the brain and act as allosteric modulators of the $GABA_A$ receptor. Since their synthesis is itself regulated by stress and by anxiogenic stimuli, targeting the neurosteroid-$GABA_A$ receptor axis represents an attractive target for the modulation of anxiety (Philippe Nuss,2015).

NEUROTRANSMITTERS INFLUENCED BY THE VARIOUS DRUGS

Perhaps the greatest practical application for the discovery and detailed understanding of how neurotransmitters function has been the development of drugs that impact chemical transmission. These drugs are capable of changing the effects of neurotransmitters, which can alleviate the symptoms of some diseases.

- **Agonists vs Antagonists:** Some drugs are known as agonists and function by increasing the effects of specific neurotransmitters. Other drugs and referred to as antagonists and act to block the effects of neurotransmission.
- **Direct vs Indirect Effects:** These neuro-acting drugs can be further broken down based on whether they have a direct or indirect effect. Those that have a direct effect work by mimicking the neurotransmitters because they are very similar in chemical structure. Those that have an indirect impact work by acting on the synaptic receptors.

Drugs that can influence neurotransmission include medications used to treat illness including depression and anxiety, such as SSRIs, tricyclic antidepressants, and benzodiazepines. Illicit drugs such as heroin, cocaine, and marijuana also have an effect on neurotransmission. Heroin acts as a direct-acting agonist, mimicking the brain's natural opioids enough to stimulate their associated receptors. Cocaine is an example of an indirect-acting drug that influences the transmission of dopamine.

Neurotransmitters regulate each other's release, and weak consistent imbalances in this mutual regulation were linked to temperament in healthy people (*Bittigau,1997;* Netter, 1991; *Trofimova, 2016;* Cloninger, 1993; *Trofimova IN, 2016)*.

CONCLUSION

Strong imbalances or disruptions to neurotransmitter systems have been associated with many diseases and mental disorders. These include Parkinson's, depression, insomnia, Attention Deficit Hyperactivity Disorder (ADHD), anxiety, memory loss, dramatic changes in weight and addictions. Chronic physical or emotional stress can be a contributor to neurotransmitter system changes. Genetics also plays a role in neurotransmitter activities. Apart from recreational use, medications that directly and indirectly interact one or more transmitter or its receptor are commonly prescribed for psychiatric and psychological issues.

REFERENCES

Bittigau, P., & Ikonomidou, C. (1997). Glutamate in neurologic diseases. *Journal of Child Neurology*, *12*(8), 471–485. doi:10.1177/088307389701200802 PMID:9430311

Booij, L., van der Does, A. J., Haffmans, P. M., Spinhoven, P., & McNally, R. J. (2005). Acute tryptophan depletion as a model of depressive relapse: Behavioural specificity and ethical considerations. *The British Journal of Psychiatry*, *187*(2), 148–154. doi:10.1192/bjp.187.2.148 PMID:16055826

Cloninger, C. R., Svrakic, D. M., & Przybeck, T. R. (1993). A psychobiological model of temperament and character. *Archives of General Psychiatry*, *50*(12), 975–990. doi:10.1001/archpsyc.1993.01820240059008 PMID:8250684

Ellis, J. R., Ellis, K. A., Bartholomeusz, C. F., Harrison, B. J., Wesnes, K. A., Erskine, F. F., ... Nathan, P. J. (2006). Muscarinic and nicotinic receptors synergistically modulate working memory and attention in humans. *The International Journal of Neuropsychopharmacology*, *9*(02), 175–189. doi:10.1017/S1461145705005407 PMID:15877932

Everitt, B. J., & Robbins, T. W. (2005). Neural systems of reinforcement for drug addiction: From actions to habits to compulsion. *Nature Neuroscience*, *8*(11), 1481–1489. doi:10.1038/nn1579 PMID:16251991

Fischbach, G. D. (2006). Schizophrenia: Signals from the other side. *Nature Medicine*, *12*(7), 734–735. doi:10.1038/nm0706-734 PMID:16829916

Gross, L. (2006). Supporting" players take the lead in protecting the overstimulated brain. *PLoS Biology*, *4*(11).

Healy, D. (2015). Serotonin and depression. *British Medical Journal*, ▪▪▪, 350.

Kodirov, S. A., Takizawa, S., Joseph, J., Kandel, E. R., Shumyatsky, G. P., & Bolshakov, V. Y. (2006). Synaptically released zinc gates long-term potentiation in fear conditioning pathways. *Proceedings of the National Academy of Sciences of the United States of America*, *103*(41), 15218–15223. doi:10.1073/pnas.0607131103 PMID:17005717

Kuffler, S. W., & Nicholls, J. G. (1976). *From Neuron to Brain*. Sunderland, MA: Sinauer Associates.

Laruelle, M., Kegeles, L. S., & Abi-Dargham, A. (2003). Glutamate, dopamine, and schizophrenia: From pathophysiology to treatment. *Annals of the New York Academy of Sciences*, *1003*(1), 138–158. doi:10.1196/annals.1300.063 PMID:14684442

Mesulam, M. (2004). The cholinergic lesion of Alzheimer's disease: Pivotal factor or side show? *Learning & Memory*, *11*(1), 43–49. doi:10.1101/lm.69204 PMID:14747516

Miller, G. M. (2011). The emerging role of trace amine-associated receptor 1 in the functional regulation of monoamine transporters and dopaminergic activity. *Journal of Neurochemistry*, *116*(2), 164–176. doi:10.1111/j.1471-4159.2010.07109.x PMID:21073468

Netter, P. (1991). Biochemical variables in the study of temperament. In J. Strelau & A. Angleitner (Eds.), *Explorations in temperament: International perspectives on theory and measurement 147–161*. New York: Plenum Press. doi:10.1007/978-1-4899-0643-4_10

NIMH. (2014). *Brain Basics*. U.S. National Institutes of Health.

Nuss, P. (2015). Anxiety disorders and GABA neurotransmission: A disturbance of modulation. *Neuropsychiatric Disease and Treatment*, *11*, 165–175. PMID:25653526

Raven, P. H., & Johnson, G. B. (1992). *Biology* (3rd ed.). St. Louis, MO: Mosby Year Book.

Robinson, T. E., & Becker, J. B. (1986). Enduring changes in brain and behavior produced by chronic amphetamine administration: A review and evaluation of animal models of amphetamine psychosis. *Brain Research*, *396*(2), 157–198. doi:10.1016/0165-0173(86)90002-0 PMID:3527341

Sarter, M., Bruno, J., & Parikh, V. (2007). Abnormal Neurotransmitter Release Underlying Behavioral and Cognitive Disorders: Toward Concepts of Dynamic and Function-Specific Dysregulation. *Neuropsychopharmacol*, *32*(7), 1452–1461. doi:10.1038j.npp.1301285 PMID:17164812

Sarter, M., & Parikh, V. (2005). Choline transporters, cholinergic transmission and cognition. *Nature Reviews. Neuroscience*, *6*(1), 48–56. doi:10.1038/nrn1588 PMID:15611726

Sarter, M. F., & Bruno, J. P. (1994). Cognitive functions of cortical ACh: Lessons from studies on trans-synaptic modulation of activated efflux. *Trends in Neurosciences*, *17*(6), 217–221. doi:10.1016/0166-2236(94)90001-9 PMID:7521080

Schacter, D.L., Gilbert, D.T., & Wegner, D.M. (2009). *Psychology*. Academic Press.

Snyder, S. H., & Innis, R. B. (1979). Peptide neurotransmitters. *Annual Review of Biochemistry*, *48*(1), 755–782. doi:10.1146/annurev.bi.48.070179.003543 PMID:38738

Trofimova, I., & Robbins, T. W. (2016). Temperament and arousal systems: A new synthesis of differential psychology and functional neurochemistry. *Neuroscience and Biobehavioral Reviews*, *64*, 382–402. doi:10.1016/j.neubiorev.2016.03.008 PMID:26969100

Trofimova, I. N. (2016). The interlocking between functional aspects of activities and a neurochemical model of adult temperament. In M. C. Arnold (Ed.), *Temperaments: Individual Differences, Social and Environmental Influences and Impact on Quality of Life* (pp. 77–147). New York: Nova Science Publishers, Inc.

Van Praag, H. M. (1981). Management of depression with serotonin precursors. *Biological Psychiatry*, *16*(3), 291–310. PMID:6164407

Whishaw, B. K., & Ian, Q. (2014). An introduction to brain and behavior (4th ed.). New York, NY: Worth Publishers.

Yang, J. L., Sykora, P., Wilson, D. M. III, Mattson, M. P., & Bohr, V. A. (2011). The excitatory neurotransmitter glutamate stimulates DNA repair to increase neuronal resiliency. *Mechanisms of Ageing and Development*, *132*(8-9), 405–411. doi:10.1016/j.mad.2011.06.005 PMID:21729715

KEY TERMS AND DEFINITIONS

ACh: In the autonomic nervous system, acetylcholine (ACh) is the neurotransmitter in the preganglionic sympathetic and parasympathetic neurons. ACh is also the neurotransmitter at the adrenal medulla and serves as the neurotransmitter at all the parasympathetic innervated organs.

Alzheimer: Alzheimer's disease is a progressive disorder that causes brain cells to waste away (degenerate) and dies. Alzheimer's disease is the most common cause of dementia a continuous decline in thinking, behavioral and social skills that disrupts a person's ability to function independently.

Dopamine: In the brain, dopamine functions as a neurotransmitter—a chemical released by neurons (nerve cells) to send signals to other nerve cells. The brain includes several distinct dopamine pathways, one of which plays a major role in the motivational component of reward-motivated behavior.

GABA: GABA (gamma-aminobutyric acid) is an inhibitory neurotransmitter that is very widely distributed in the neurons of the cortex. GABA contributes to motor control, vision, and many other cortical functions. It also regulates anxiety.

Memory: Memory is the faculty of the brain by which data or information is encoded, stored, and retrieved when needed. It is the retention of information over time for the purpose of influencing future action.

Parkinson: Parkinson's disease is a progressive nervous system disorder that affects movement. Symptoms start gradually, sometimes starting with a barely noticeable tremor in just one hand. Tremors are common, but the disorder also commonly causes stiffness or slowing of movement.

Serotonin: Serotonin or 5-hydroxytryptamine is a monoamine neurotransmitter. It has a popular image as a contributor to feelings of well-being and happiness.

Chapter 7 Foundations of Neuropsychology

Barre Vijaya Prasad
https://orcid.org/0000-0003-2332-8220
All India Institute of Medical Sciences (AIIMS), India & Dharwad Institute of Mental Health and Neurosciences, India

Sumit Kumar
Jawaharlal Nehru University, India

Shazia W Ali
Sher-I-Kashmir Institute of Medical Sciences (SKIMS), India

ABSTRACT

Neuropsychology is one of the branches in neuroscience (or sub-branch of psychology) that seeks to understand the brain structure and further relates it with specific human behavior and psychological processes. Neuropsychology's central aim is to understand how the brain produces behavior, mental processes, and cognition and then applies it for diagnosis and treatment of different behavioral and cognitive defects in various neurological disorders. Neuropsychology is a field that draws information and material from neurology, cognitive psychology, and psychiatry and also act as a junction point for investigators belonging to related disciplines. Its central aim is to understand how brain functioning influences behavior and cognition. Further, the chapter discusses the relationship between structure and function of the brain, the rise of neuropsychology as a distinct discipline, logic of cerebral organization, and localization and lateralization of functions. Finally, the authors had also explain the approaches and methodologies of clinical and cognitive neuropsychologists.

DOI: 10.4018/978-1-7998-2860-0.ch007

CHAPTER OUTLINE

- Introduction
- Relationship between structure and function of the brain
- The rise of neuropsychology as a distinct discipline
- Logic of cerebral organization
- Localization and lateralization of functions
- Approaches and methodologies of clinical and cognitive neuropsychologists

INTRODUCTION

Neuropsychology is one of the branch in neuroscience (or sub-branch of psychology) that seeks to understand the brain structure and further relates it with specific human behavior and psychological processes. Neuropsychology central aim is to understand how the brain produces behavior, mental processes, and cognition and then apply it for diagnosis and treatment of different behavioral and cognitive defects in various neurological disorders. Neuropsychology is a field that draws information and material from neurology, cognitive psychology, and psychiatry and also act as a junction point for investigators belongs to related discipline. In general, it shares both concept and concern with neuropsychiatry and behavioral neurology. Neuropsychology is consisting of both experimental and clinical field of psychology. Its central aim to understand how brain functioning influences behavior and cognition. The main focus of classical neurology was the pathology of the nervous system, while classical psychology was primarily devoted to the analysis of human behavior. Thus neuropsychology strives for to how the brain relates with the mind through the study of neurological patients. Hence it shares fundamental with both neuropsychiatry and behavioral neurology. Originally neuropsychology term has been used for lesion studies and electrical activity (single neuron or a group of the neuron) in humans and animals.

RELATIONSHIP BETWEEN STRUCTURE AND FUNCTION OF THE BRAIN

The study of the brain is relatively a newer field as pre-medieval peoples held the impression that the heart is central point for thinking rather than brain. The focus of European renaissance was the revival of classical art, literature, philosophy, music, architecture, and also placed a particular emphasis on humanism. During that period, there was a renewed interest to understand the human body and its association with

Figure 1.

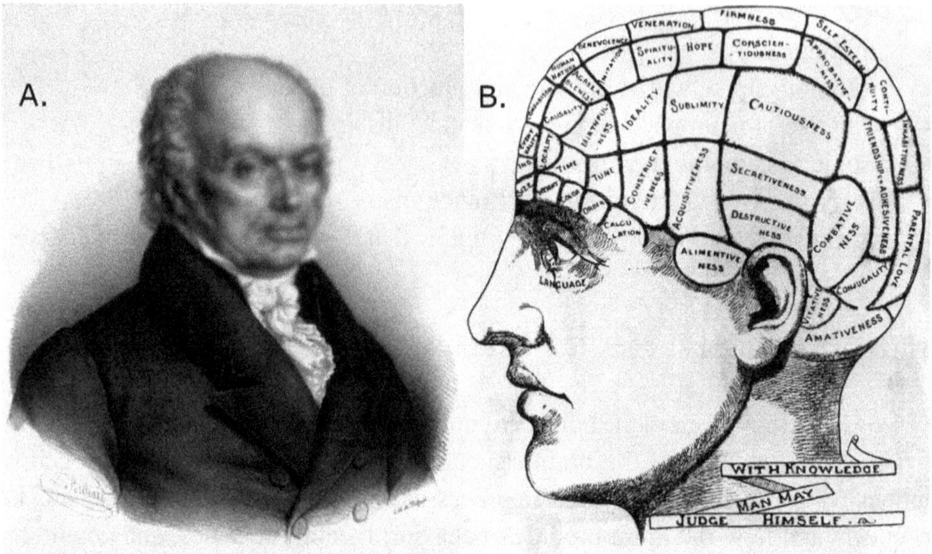

the mind. Although different philosopher tried to associate the different behavior function with various brain parts but failed miserably due to lack of strong foundation. They studied the ventricles or cavities surrounding the brain, corpus callosum, pineal and pituitary glands, although their exact function was still a mystery in that time. Descartes in 1664, first time documented the pineal gland as a junction point of all body sensory inputs and provide a non-physical sense of awareness. The first real effort to unravel the brain-behavior phenomena was made by an Austrian physician named Gall, and his student Spurzheim (Figure 1a). Gall agreed that the brain is the main control center of the body, consisting of 27 compartments, and further differentiating it into grey and white matter. He also associated the aphasia or impaired language production with frontal damage. Spurzheim had a deep interest in brain anatomy. He divided the brain into the different part with assign function, its connections with the spinal cord, and furthers its involvement in muscles control.

The seminal work of Gall and Spurzheim lead to the birth of "science of phrenology", studying the individual personality and other features based on skull physical size and shape (Figure 1b). However, soon phrenology was demised due to lack of any proof and further disproving it through experiments by Pierre Flourens. Then, French physicians, Bouillaud, and Dax had independently described a handful of patients they had lost the speech after brain damage.

Brain or the encephalon is a rounded walnut, grayish in color, weighting 1.2–1.4 kg or 2% of total body weight. It is surrounded by a layer of tissue called the meninges. It contains billions of neural networks into many specialized areas that work together to regulate the different body process and also produces a trait called behavior. It is possibly the most complex organ to present within the human body. Without brain, our bodies simply would not function, and thus it is crucial to have a proper understanding of its structure and function, pharmacology associated with mental illness and implications of disease and diagnosis. Brain is divided into multiple regions (Figure 2a). The grooves are called fissures if extend deep into the brain or sulci if they are shallower than fissures and the bumps that we see are called Gyri. It serves as markers to identify regions of the brain.

Human brain anatomical regions are ordered into following part anatomically:

- **Brainstem:** The brainstem is the oldest and fundamental part of the brain, beginning where the spinal cord swells and enters the skull (Figure 2a). It is responsible for automatic survival functions. The medulla is the base of the brainstem that controls heartbeat and breathing. While pons helps with movement and facial expression. Reticular Formation is a nerve network present in the brainstem, plays an essential role in controlling arousal.
  The thalamus is the brain sensory switchboard, located on top of the brainstem. It directs messages to the sensory areas present in the cortex and transmits replies to the cerebellum and medulla. It receives information for all of the senses except for the smell.
- **Cerebellum:** The cerebellum is called the "little brain" and is attached to the rear of the brainstem (Figure 2a). It helps to coordinate voluntary movements and control balance. It also plays a part in memory, emotion regulation, timing, emotional modulation, and sensory discrimination.
- **The Limbic System:** The Limbic System is a doughnut-shaped structure present in neural structures at the border of the brainstem and cerebrum, associated with strong emotions such as fear, aggression, and drives for food and sex. It includes the hippocampus, amygdala, and hypothalamus (Figure 2a).
- **Hippocampus:** The hippocampus processes memories (Figure 2a).
- **Amygdala:** The amygdala consists of two almond-shaped neural clusters linked to the emotions of fear and anger (Figure 2a).
- **Hypothalamus:** The hypothalamus lies below (hypo) the thalamus. It directs several maintenance activities like thirst, fatigue, sleep, eating, drinking, body temperature, and control of emotions. It also controls the crucial aspects of parenting and attachment behaviors. It helps control the endocrine system by giving directions to the pituitary gland (Figure 2a).

Figure 2.

A.

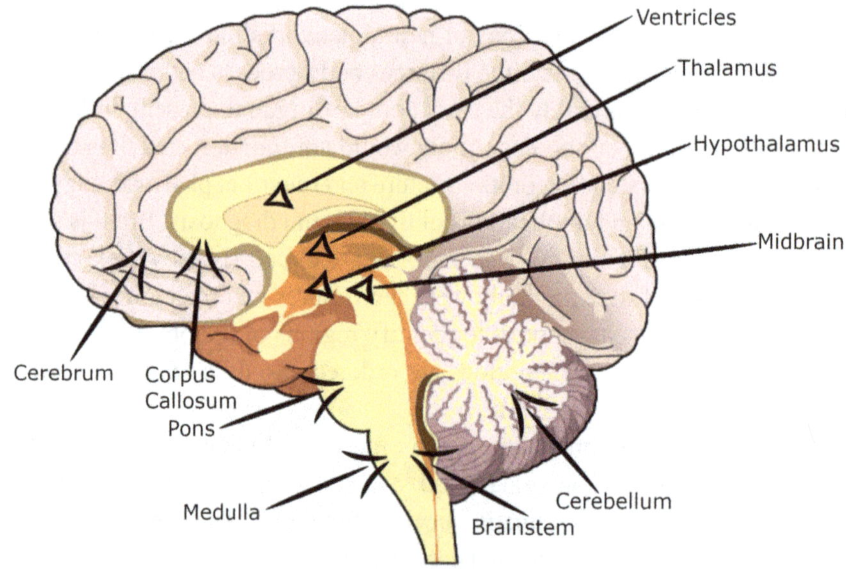

B.

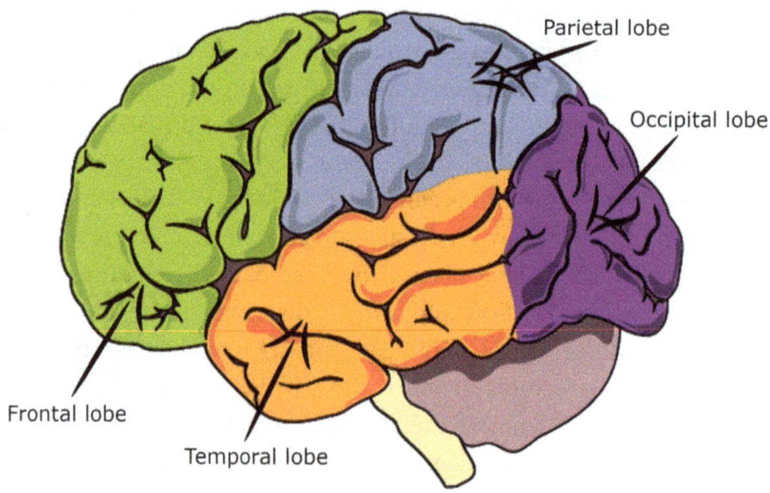

- **The Cerebral Cortex:** The cerebral cortex is an outermost layer of brain cells, wrinkled around the inner part of brain tissue and divided into two hemispheres: right and left which are connected by bundle of fibers called the corpus callosum that provide a path to communicate one side to the other (Fig. 2a). The two hemispheres of the brain are separated by cutting the connecting fibers to prevent seizures in patients with severe epilepsy. Each side controls the opposite side of the body. Left hemisphere in the majority of people is the dominant hemisphere- responsible for the production of language, mathematical ability, problem-solving, and logic. Right hemisphere thought to be responsible for creative thinking and spatial ability.

Each hemisphere is further divided into four main lobes: the frontal lobe, parietal lobe, temporal lobe, and occipital lobe. It evolved recently and cover around 80% of the brain part. The association areas integrate sensory information and stored memories. The "uncommitted" or association areas of the cortex increase with intelligence. It is responsible for the cognition, mental imagery, language processing, and several other and voluntary movements. When looking at the brain, what makes it distinctive is the numerous folds that give it its wrinkled appearance. This folding present in brain tissue allows for greater amount of cerebral surface area (approximately two-thirds of the total cerebral surface area is located in the depths of these folds) to be confined within the small space of the skull, leading to more information being relayed throughout areas of the brain. Each brain hemisphere is segregated into four lobes, and each one is separated by prominent fissures (Figure 2b).

- **Frontal Lobe:** It is located in the front of both cerebral hemispheres and the largest lobes of the brain (Figure 2b). It consists primary motor cortex, represents particular areas of the body, for example, the middle area of the cortex controls the legs, while the lateral region is for the muscles of the face and rest area located between both of these areas, which is also largest area, control arm, and hands. The posterior parietal cortex, responsible for transforming visual information into motor commands
    - The pre-motor cortex, responsible for motor guidance of movement and control of proximal and trunk muscles of the body.
    - The supplementary motor area (or SMA) responsible for planning and coordination of complex movements such as those requiring two hands.

The frontal lobes of brain cerebral cortex are also thought to be involved in complex functioning on the brain including personality, judgment, insight, reasoning, skeletal movement, problem-solving, abstract thinking, speech control, and self-evaluation termed executive functions. The frontal lobe also has a role in integrating longer non-task based working memory, especially the ability to plan and initiate activity that fit into socially acceptable behavior.

- **Parietal Lobe:** It is located behind the frontal lobe and also known as somatosensory cortex as it receives general sensory information and initial reception of tactile (touch, pain, and temperature information, etc.) and proprioceptive (sense of position) information (Fig. 2b). The other leading role of the parietal lobes are complicated form of spatial orientation and perception, and the comprehension of language function and the ability to recognize objects by touch, calculate, write, recognize fingers of opposite hands and organize spatial directions. The posterior areas of the parietal lobes (through the dorsal stream of the visual cortex) appear to link visual and somatosensory information together. Damage to these areas produces the neglect of entire spaces of sensory information, for example, only eating from one side of a plate and other sensory deficits.

- **Temporal Lobe:** It is located at each side of the brain (Figure 2b). It involved in receiving and processing auditory information, higher orders visual information, complex aspects of memory and language. Wernicke's area is present in temporal lobe responsible for comprehension of speech and closely linked with Broca's area to produce speech. Damage to this area is associated with diminished thiamine levels due to alcohol abuse called delirium.
- **Occipital Lobe:** It is present at the rearmost portion of the brain (Figure 2b). The primary visual cortex present in the occipital lobe, which receives raw sensory information from the retina processes information on object color cues and facial recognition and is also involved in the perceiving motion. It also contains corpus callosum, the fiber bundle in the brain that connects the two hemispheres together and allows the transmission of information between the two hemispheres. Damage to the visual cortex results in cortical blindness (all structures to see are intact but the person cannot receive the input from the sensors).

THE RISE OF NEUROPSYCHOLOGY AS A DISTINCT DISCIPLINE

Neuropsychology is a relatively new discipline emerged within psychology. However, the history of Neuropsychology can be traced back to the time of ancient Egypt priest Imhotep, describing brain, trauma, abnormalities, and remedies. However, they took heart as the seat of the soul rather than brain. This belief was further reinforced by Aristotle and Hippocrates and persisted till 17th century. Since then neuropsychology was much part of either behaviorism or psychology till the 1950s. Neuropsychology seeks to understand the relationship between the brain and behavior, that is, it attempts to explain the way in which the activity of the brain is expressed in observable behavior. It retains a degree of distinctiveness that distinguishes it from other related areas like for example behavioral neurology, biological psychology, clinical neuropsychology, cognitive neuropsychiatry, cognitive neuropsychology, cognitive neuroscience, cognitive psychology, comparative neuropsychology, neurology and neuroscience. A psychology without any reference to physiology can hardly be complete. The operation of the brain is relevant to human conduct, and the understanding of how the brain relates to behavior may make a significant contribution to understanding how other, more purely psychological, factors operate in directing behavior. Just how the brain deals with intelligent and complex human functions is, in any case, an important subject of investigation in its own right, and one that has an immediate relevance for those with brain injuries and diseases, as well as a wider relevance for medical practice.

The term neuropsychology first used by Sir William Osler in 1913. It also entered psychology nomenclature after use by Karl Lashley in 1936. The term "neuropsychology" was not known till 1950s and first time find mention without definition in book named "The organisation of behaviour: A neuropsychological theory", authored by Donald Hebb's in 1949 . From 1950s-70s there was not much development in this field except the establishment of the International Neuropsychological Society in 1967. In 1980, Kolb and Whishaw made the first sincere attempt to define neuropsychology in his book "Fundamentals of Human Neuropsychology" and Lezak's Neuropsychological assessment in 1980-1983 since then it became popular, widely used and appeared in various researches. From 1990s, neuropsychology has seen a paradigm shift in neuropsychology by leveraging its position at the interface of basic biology and clinical science and then its integration with information science, leading to healthcare reform. Further various technological advancement in visualizing of brain structure and function by fMRI and integrating it with informatics, NIH Human Connectome Project, Human genome project to The Human Phenome Project under the aegis of Consortium for Neuropsychiatric Phenomics are leading the neuropsychology from the forefront and giving it a new independent identity out of psychology.

However, it would be fallacious for us to argue that neuropsychology is a unified discipline. It is broadly divided into two domains: clinical and cognitive neuropsychology. Early practitioners and researcher of clinical neuropsychology were tended to work in neurology clinics with a focus on to understand the behavioral impairments associated with discrete brain lesions and trauma. While letter study the impairments to psychological processes originate from the altered brain, lesions, and trauma. Put differently the clinical neuropsychology damaged brain to psychological dysfunction, while cognitive neuropsychology goes from brain dysfunction to offering explanation about observed behavior.

ORIGIN OF BRAIN HYPOTHESIS

Localization of brain function has been the focus of philosophers, physiologists, and psychologists for many centuries. Around 400 BC, Hippocrates attempted to correlate his behavioral observations with what he knew about anatomical localization; this was conjecture because he was legally and socially prohibited from dissecting the human body, especially the cranium. Later, Aristotle surmised that the heart was the seat of the mind. Almost 600 years after Hippocrates, Galen shifted the site of the mind to the brain. Clarification of overall mind function was later offered by Descartes who suggested that the soul was localized in the pineal gland. In 1810, Gall described cortical localization of function through the concept of phrenology.

Flourens and Broca's introduced more accurate observations of brain function during the mid 19th century, forcing physiologists to research localization of function more systematically and with more precisely measurement tools. This research was informed by early recordings of brain function or dysfunction which can be traced to at least the 17th century (Gibson, 1962) when several cases traumatic brain injury were documented. Precise experimental, rather than observational, analysis began with the electrical stimulation work of Fritsch and Hietzig in 1790. New scientific techniques were introduced the study of brain function by one of Catell's students, Sheperd Franz, during the early part of this century while in Washington, D.C., Franz taught Karl S. Lashley who is in turn, advanced the understanding of brain-behavior relationships as well as the theory of equipotentiality. During the mid 20th century, Nobel laureate Roger Sperry and colleagues extended this earlier work by developing procedures for experimentally examining disconnection syndromes. For many centuries, the brain was thought useless and was often discarded during burial processes and autopsies. As the field of medicine developed its understanding of human anatomy and physiology, different theories were developed as to why the body functioned the way it did. Many times, bodily functions were approached from a religious point of view and abnormalities were blamed on bad spirits and the gods. The brain has not always been considered the center of the functioning body. It has taken hundreds of years to develop our understanding of the brain and how it affects our behaviors (see the Figure 3).

LOGIC OF CEREBRAL ORGANIZATION

The cerebral cortex is the most outermost layer of neural tissue and synonymous with intelligence. It comprises "primitive" regions i.e., hippocampus and the "newly" evolved areas such as the six-layered "neocortex". It is only present in mammals and most prominently developed in humans. So now the question is when did these components arise and then how did they evolve into higher-order supracomplex function such as cognition, emotion, and consciousness. The neocortex is the youngest part of the brain consisting of 76% of entire brain size. The six-layered neocortex evolutionarily derived from three-layer cortex present in all non-avian modern reptiles. The three-layer cortex is evolutionary conserved, consists hippocampus, etc. and believed to be transformed into neocortex in a mammal during the transition from Triassic and Jurassic periods. The brain is built to work on logical way; therefore structurally or anatomically it must be organized to look logical in different organism. Organism brain size is independent to body weight, like blue whales (Balaenopteramusculus) with a bodyweight of 100,000 kg have 6.9 kg brain, while Indian elephant (Elephasindicus) possess 6 kg of the brain with a 20-fold lighter body (5,000 kg) than blue whale. Further, even with variable brain size, cerebral cortex size does not change much, indicating similar origination and architecture. However, the minor change in cortical thickness is strongly associated with the intelligence level.

Figure 3. The growth rate of hominid cranial capacity began to rapidly increase starting around two million years ago

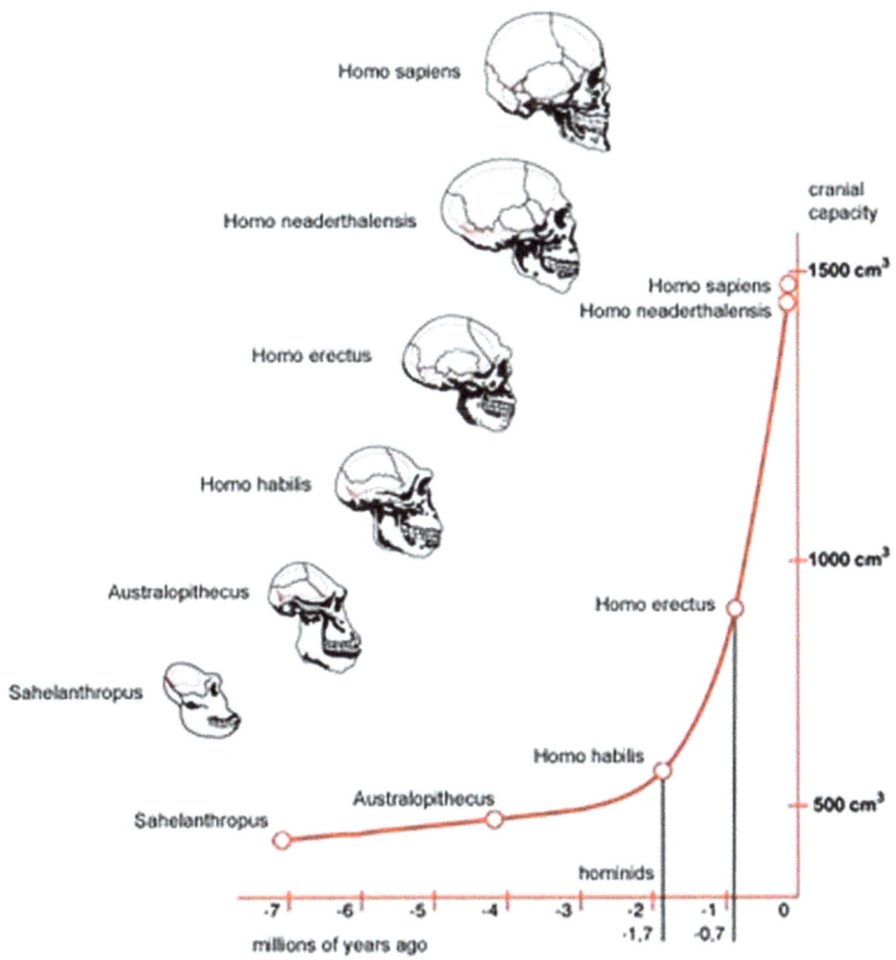

It is not just brain size, but the cortical folding of the brain also increases with the organism complexity. The more convoluted brain surface offers the larger cortical surface area and thus allows the cortical expansion with evolutionary advancement. However, with increasing cortical thickness, it is difficult for the brain to fold; therefore, it seems that evolution has made a fine balance between cortical thickness and cortical convolution.

Neocortex consisted of six layers, representing the highest level of functioning and cognitive ability. Neocortex expansion or development in any species requires similar changes in the brain to support the expansion. The neocortex expansion is associated with greater cooperation and competition in the organism, and thus it offers greater voluntary control over organism social behaviors without changing brain size and thus leading to increased social harmony, cooperation, and wellbeing.

The neurons are the fundamental building blocks of the brain; therefore, they must be organized in such a way that saves the energy resources. One hypothesis has been advanced to explain functional placement of neuron is called 'Minimal Wiring Hypothesis' meaning neurons involved in more frequent task or coordinated movement are localized closer to avoid delay, and energetically expensive long-range neuronal connections.

LOCALIZATION AND LATERALIZATION OF BRAIN FUNCTION

- **Localization:** The cerebral cortex is divided into many different areas, each of which is closely associated with specific mental and behavioural functions. This division of the cortex is based on difference is detailed microscopic structure, anatomical connections, and functional properties (what it does) among different areas. Localization of function is the idea that certain functions (e.g. language, memory, etc.) have certain locations or areas within the brain. This idea has been supported by recent neuroimaging studies, but was also examined much earlier, typically using case studies. One such case study is that of Phineas Gage, who in 1848 while working on a rail line, experienced a drastic accident in which a piece of iron went through his skull. Although Gage survived this ordeal, he did experience a change in personality, such as loss of inhibition and anger. This change provided evidence to support the theory of localisation of brain function, as it was believed that the area the iron stake damaged was responsible for personality. There are four key areas that you need to be aware of: motor, somatosensory, visual and auditory areas (see Figure 4).
- **Motor Area:** The motor area is located in the frontal lobe and is responsible for voluntary movements by sending signals to the muscles in the body. Hitzig

and Fritsch (1870) first discovered that different muscles are coordinated by different areas of the motor cortex by electrically stimulating the motor area of dogs. This resulted in muscular contractions in different areas of the body depending on where the probe was inserted. The regions of the motor area are arranged in a logical order, for example, the region that controls finger movement is located next to the region that controls the hand and arm and so on.

- **Somatosensory Area:** The somatosensory area is located in the parietal lobe and receives incoming sensory information from the skin to produce sensations related to pressure, pain, temperature, etc. Different parts of the somatosensory area receive messages from different locations of the body and this area of the brain is highly adaptable, with Braille readers having larger areas in the somatosensory area for their fingertips compared to normal sighted participants.
- **Visual Area:** At the back of the brain, in the occipital lobe is the visual area, which receives and processes visual information. Information from the right-hand side visual field is processed in the left hemisphere, and information from the left-hand side visual field is processed in the right hemisphere. The visual area contains different parts that process different types of information including colour, shape or movement.
- **Auditory Area:** The auditory area is located in the temporal lobe and is responsible for analysing and processing acoustic information. Information from the left ear goes primarily to the right hemisphere and information from the right ear goes *primarily* to the left hemisphere. The auditory area contains different parts, and the primary auditory area is involved in processing simple features of sound, including volume, tempo and pitch.
- **What we Know Brodmann's Areas:** Korbinian Brodmann was a German neuroanatomist in the late 1800s and early 1900s. He is best known for dividing the cerebral cortex of the brain into 52 different regions based on their histological properties. His major study in cytoarchitecture was published in 1909 (Comparative Localization Studies in the Brain Cortex, its Fundamentals Represented on the Basis of its Cellular Architecture). In this work, he analyzed the brains of humans, nonhuman primates, and other species (keep in mind that having the same Brodmann numbers across species does not indicate that the region of the brain performs the same function.) By taking different parts of brains and processing them with a Nissl stain, he could examine the types of cells contained within. He looked at general cytoarchitecture as well as internal circuitry of how cells were connected. One of Brodmann's theories was that each of these different parts of the

Table 1. Brodmann's areas

Areas 1, 2, 3	Primary somatosensory cortex (postcentral gyrus)
Area 4	Primary motor cortex (precentral gyrus)
Area 5	Somatosensory association cortex
Area 6	Premotor and supplementary motor cortex
Area 9	Dorsolateral/anterior prefrontal cortex (motor planning, and organization)
Area 10	Anterior prefrontal cortex (memory retrieval)
Area 17	Primary visual cortex
Area 22	Primary auditory cortex
Area 37	Occipitotemporal (fusiform) Gyrus
Areas 22, 39, 40	Wernicke's area (language comprehension)
Areas 44, 45	Broca's area (motor speech programming)

brain performed a different function, a line of thought that still persists today (Table 1).

- **Areas 1,2,3:** Primary somatosensory cortex (or Postcentral gyrus) – this is numbered rostral to caudal as 3,1,2. This region is associated with several senses, such as the ones for:
- Localization of touch, temperature, vibration, pain
- Sensory perception (two-point discrimination, proprioception, etc.) especially the legs, trunk, arms, hands, face and lips
- Skilled and coordinated orofacial movement (i.e. whistling)
- Motor learning
- **Area 4:** This are is also known as the primary motor cortex (or Precentral gyrus) (may possibly include part of Area 6). It is responsible for executing motor movements, which includes contralateral finger/hand/wrist or orofacial movements, learned motor sequences, breathing control, and voluntary blinking.
- **Area 5:** Somatosensory association cortex – Brodmann area 5 is part of the parietal cortex. Associated functions include:
    - Stereopsis
    - Line bisection judgments
    - Processing chaotic patterns
    - Using spatial imagery in deductive reasoning
    - Motor execution
    - Bimanual manipulation
    - Working memory

Figure 4. Motor and sensory regions of the cerebral cortex

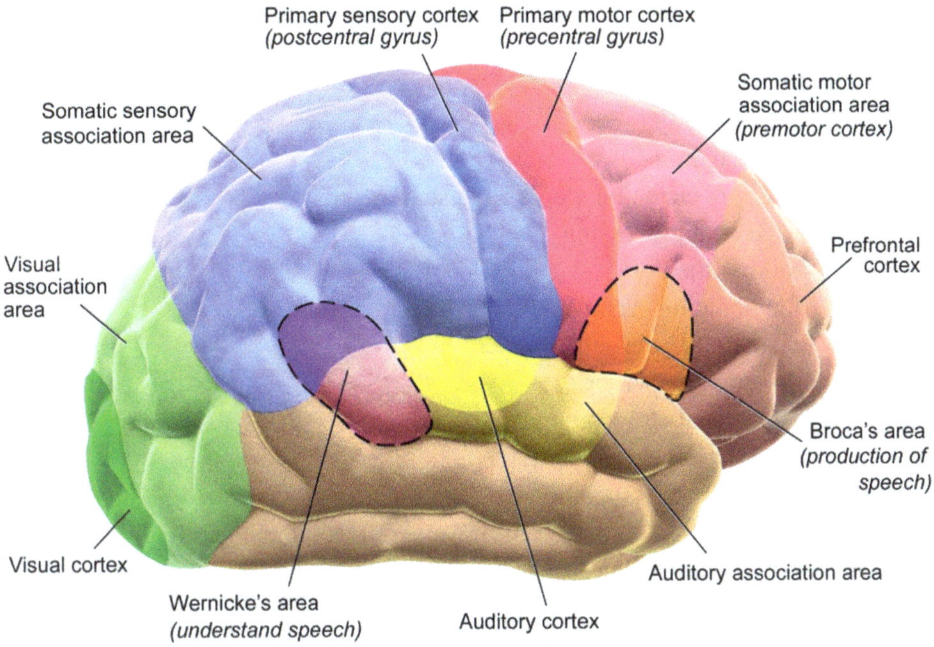

- Language processing
- Visuomotor attention
- Pain perception
- Tactile localization
- Saccadic eye movement

- **Area 6:** Premotor and Supplementary Motor Cortex – this region is critical for the sensory guidance of movement and control of proximal and trunk muscles, and contributes to the planning of complex and coordinated motor movements. This area plays a large role in motor, language, and memory functions, including: Premotor and Supplementary Motor Cortex – this region is critical for the sensory guidance of movement and control of proximal and trunk muscles, and contributes to the planning of complex and coordinated motor movements. This area plays a large role in motor, language, and memory functions, including:
    - Motor sequencing/planning
    - Laughter/smiling
    - Interlimb coordination
    - Movement learning and initiation
    - Motor imagery

  - Speech motor programming
  - Language processing and switching
  - Speech perception
  - Object naming
  - Lip-reading
  - Word retrieval
  - Lexical decision on words and pseudowords
  - Syntactical processing
  - Working memory
  - Mnemonic rehearsal
  - Episodic long-term memory
  - Topographic memory
  - Visuospatial and visuomotor attention
  - Attention to human voices

  Note: The "Motor Association Cortex" includes areas 6, 8, 44, and 45. This association cortex is involved in movement throughout the body, including motor speech movements.
- **Area 9:** Dorsolateral/anterior prefrontal cortex (DLPFC) – This region is the highest cortical area responsible for motor planning, organization, and regulation, and sustaining attention and working memory. The DLPFC plays an important role in:
  - Integrating sensory and mnemonic information
  - Regulation of intellectual function and action
  - The act of deception and lying

  In sum, all complex mental activities require crosstalk between cortical and subcortical circuits that are connected to the DLPFC.
- **Area 10:** Anterior prefrontal cortex – involved in strategic processes of memory retrieval and executive functions, such as:
  - Anterior prefrontal cortex – involved in strategic processes of memory retrieval and executive functions, such as
  - Reasoning
  - Task flexibility
  - Problem solving
  - Planning
  - Execution
  - Working memory
  - Processing emotional stimuli
  - Inferential reasoning
  - Decision making
  - Calculation of numerical processes

There is a significant role of Areas 9 and 10 in memory encoding and retrieval, and area 10 is thought to control and manipulate event and time-based prospective memory (metamemory), and allow "intentional forgetting."
Area 10 is also involved in attending to sensory stimulation and use of language (generating sentences, word-stem completion, verbal fluency, syntactic processing, and metaphor comprehension).

- **Area 17:** Primary visual cortex (V1) – the visual cortex is located in the occipital lobe in the back of the brain, and contains a well-defined map of the spatial information required for vision.
- **Area 22:** Primary auditory cortex / Superior Temporal Gyrus (part of Wernicke's area) – this region is situated close to the external ear and involves complex language and auditory processing.
- **Area 37:** Fusiform Gyrus / Occipitotemporal Gyrus – this region is largely involved in:
    - Processing of color information
    - Face and body recognition
    - Word and number recognition
- **Areas 22, 39, 40:** Wernicke's area – the gyri that comprise this area may be larger or smaller in different people and itis responsible for speech fluency. More precisely, this area allows you to string words together in complete and sensical sentences. An easy way to remember is to think of "Wordy Wernicke".
- **Areas 44,45:** Broca's area – this region is associated with the praxis of speech (motor speech programming). This includes:
    - Being able to put together the binding elements of language
    - Selecting information among competing sources
    - Sequencing motor/expressive elements
    - Cognitive control mechanisms for syntactic processing of sentences
    - Construction of complex sentences and speech patterns

LATERALIZATION

The lateralization of brain function is the tendency for some neural functions or cognitive processes to be specialized to one side of the brain or the other. The medial longitudinal fissure separates the human brain into two distinct cerebral hemispheres, connected by the corpus callosum. Although the macrostructure of the two hemispheres appears to be almost identical, different composition of neuronal networks allows for specialized function that is different in each hemisphere. Lateralization of brain structures is based on general trends expressed in healthy patients; however, there are numerous counterexamples to each generalization; each human's brain develops differently leading to unique lateralization in individuals. This is different from specialization as lateralization refers only to the function of one structure divided between two hemispheres (see Fig 5). Specialization is much easier to observe as a trend since it has a stronger anthropological history; the best example of an established lateralization is that of Broca's and Wernicke's areas where both are often found exclusively on the left hemisphere. These areas frequently correspond to handedness, however, meaning that the localization of these areas is regularly found on the hemisphere corresponding to the dominant hand (anatomically on the opposite side) (Halpern, Gunturkun, Hopkins., & Rogers, 2005). Function

Figure 5. The human brain is divided into two hemispheres–left and right

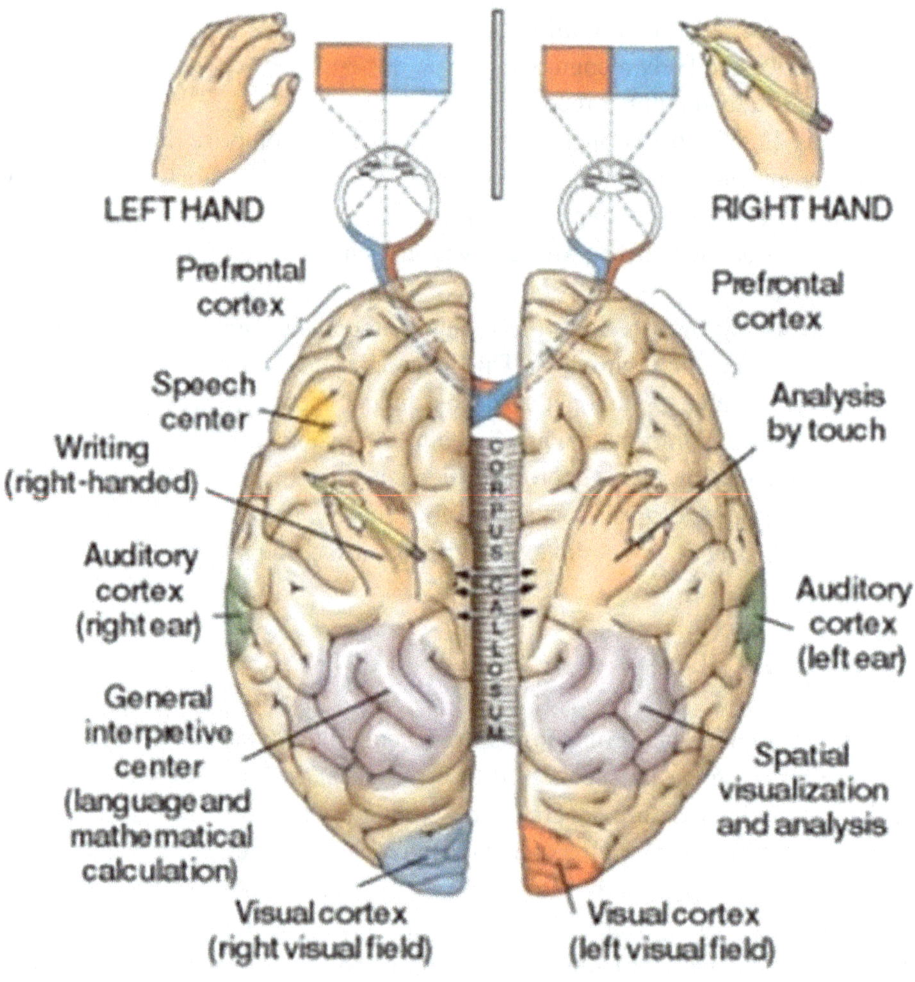

lateralization, such as semantics, intonation, accentuation, and prosody, has since been called into question and largely been found to have a neuronal basis in both hemispheres. Another example is that each hemisphere in the brain tends to represent one side of the body. In the cerebellum this is the same bodyside, but in the forebrain this is predominantly the contralateral side (Ries, Dronkers, & Knight, 2016).

FUNCTIONS OF LATERLIZATIONS

- **Language:** Language functions such as grammar, vocabulary and literal meaning are typically lateralized to the left hemisphere, especially in right-handed individuals (Taylor & Taylor, 1990) While language production is left-lateralized in up to 90% of right-handers, it is more bilateral, or even right-lateralized, in approximately 50% of left-handers (Beaumont, 2008). Broca's area and Wernicke's area areas associated with the production of speech and comprehension of speech, respectively, are located in the left cerebral hemisphere for about 95% of right-handers, but about 70% of left-handers (Richard., & Brian, 2012).
- **Sensory Processing:** The processing of basic sensory information is lateralized by being divided into left and right sides of the body or the space around the body. In vision, about half the neurons of the optic nerve from each eye cross to project to the opposite hemisphere and about half do not cross to project to the hemisphere on the same side; this means that the left side of the visual field is processed largely by the visual cortex of the right hemisphere and vice versa for the right side of the visual field. In audition, about 90% of the neurons of the auditory nerve from one ear cross to project to the auditory cortex of the opposite hemisphere.In the sense of touch, most of the neurons from the skin cross to project to the somatosensory cortex of the opposite hemisphere. Because of this functional division of the left and right sides of the body and of the space that surrounds it, the processing of information in the sensory cortices is essentially identical; that is, the processing of visual and auditory stimuli, spatial manipulation, facial perception, and artistic ability are represented bilaterally(Beaumont, 2008). Numerical estimation, comparison and online calculation depend on bilateral parietal regions (Dehaene, Spelke, Pinel, Stanescu, & Tsivkin, 1999; Dehaene, Piazza, Pinel., & Cohen, 2003). while exact calculation and fact retrieval are associated with left parietal regions, perhaps due to their ties to linguistic processing (Dehaene et al.,1999; Dehaene et al., 2003).

- **led Iain McGilchrist Value Systems:** Rather than just being a series of places where different brain modules occur, there are running similarities in the kind of function seen in each side, for instance how right-side impairment of drawing ability making patients draw the parts of the subject matter with wholly incoherent relationships, or where the kind of left-side damage seen in language impairment not damaging the patient's ability to catch the significance of intonation in speech (McGilchrist, 2009) this has led Iain McGilchrist to say that the two hemispheres as having different value systems, where the left hemisphere tends to reduce complex matters such as ethics to rules and measures, where the right hemisphere is disposed to the holistic and metaphorical (Iain McGilchrist & Shankar, 2019)
- **Clinical Conditions:** Depression is linked with a hyperactive right hemisphere, with evidence of selective involvement in "processing negative emotions, pessimistic thoughts and unconstructive thinking styles", as well as vigilance, arousal and self-reflection, and a relatively hypoactive left hemisphere, "specifically involved in processing pleasurable experiences" and "relatively more involved in decision-making processes" (Hecht, 2010). Additionally, "left hemisphere lesions result in an omissive response bias or error pattern whereas right hemisphere lesions result in a commissive response bias or error pattern (Braun, Delisle, Guimond., & Daigneault, 2009). The delusional misidentification syndromes, reduplicative paramnesia and Capgras delusion are also often the result of right hemisphere lesions (Devinsky, 2009).
- **Hemisphere Damage:** Damage to either the right or left hemisphere, and its resulting deficits provide insight into the function of the damaged area. Left hemisphere damage has many effects on language production and perception. Damage or lesions to the right hemisphere can result in a lack of emotional prosody or intonation when speaking. Right hemisphere damage also has grave effects on understanding discourse. People with damage to the right hemisphere have a reduced ability to generate inferences, comprehend and produce main concepts, and a reduced ability to manage alternative meanings. Furthermore, people with right hemisphere damage often exhibit discourse that is abrupt and perfunctory or verbose and excessive, they can also have pragmatic deficits in situations of turn taking, topic maintenance and shared knowledge. Lateral brain damage can also affect visual perceptual spatial resolution. People with left hemisphere damage may have impaired perception of high resolution, or detailed aspects of an image. People with right hemisphere damage may have impaired perception of low resolution, or big picture, aspects of an image.

- **Plasticity:** If a specific region of the brain, or even an entire hemisphere, is injured or destroyed, its functions can sometimes be assumed by a neighboring region in the same hemisphere or the corresponding region in the other hemisphere, depending upon the area damaged and the patient's age;[14] when injury interferes with pathways from one area to another, alternative (indirect) connections may develop to communicate information with detached areas, despite the inefficiencies.
- **Broca's Aphasia:** This type of aphasia is associated with lesions of Broca's area in the inferior posterior frontal lobe. It is the most generally recognized form of aphasia, and is the classical nonfluent form. The principal sign is that the patient's speech output is severely impaired. Most likely to be affected are articles, adverbs, adjectives, and other function words, so that speech tends to be reduced to nouns and verbs only. In the most extreme cases there is muteness, and in severe cases nouns may be restricted to their singular form, and verbs to the infinitive or a participle. This kind of speech is sometimes described as "telegraphic." The meaning is often evident, and words tend to be in the correct order, but it is as if speech has been reduced to its essential elements by agrammatic deletions. Short automatic phrases, of the kind common in "small talk," may well be inserted, and in fact used excessively and inappropriately. There may also be errors in the actual production of words. *Paraphasias*, as these are known, are relatively common, but are of the phonemic variety where the target word is usually identifiable. Some error in the selection or production of sound elements seems to occur, but as the word produced usually approximates to the target word, and as the context may also provide some clues, it is generally possible to tell what the patient is intending to say.

  There is, in addition, some difficulty with repetition and with naming, although prompting by sounds or context can be of significant help with the latter. However, it is probably best to see these particular difficulties as secondary to the problems with speech output, limiting the performance that might otherwise be attained. This is probably not a complete explanation, because written output may well be affected in the same way as speech, presumably by sharing certain output mechanisms, and the difficulty therefore seems to lie in language production generally rather than specifically with speech. All these problems occur, however, with language comprehension intact. Patients understand perfectly what is said to them, and read as well as they comprehend speech. They often show extreme frustration at their own inability to communicate by speech, while understanding what they read and hear, although not all of them are aware of their speech limitation.

- **Wernicke's Aphasia:** Wernicke's aphasia is the result of damage to the area of the brain that is commonly in the left hemisphere above the sylvian fissure. Damage to this area causes primarily a deficit in language comprehension. While the ability to speak fluently with normal melodic intonation is spared, the language produced by a person with Wernicke's aphasia is riddled with semantic errors, and may sound nonsensical to the listener. "Fluent" does not mean that speech is normal, and the most striking feature of patients suffering from this type of aphasia is their verbal behavior. Although they produce normal, and sometimes excessive, amounts of speech, they are generally quite unintelligible. The speech is filled by paraphasias that are in this case of a semantic nature. The patient substitutes different words for the ones that he presumably intends to utter. These may be semantically related to the target words (*red* for *green*, *table* for *chair*), but they may also be neologisms, resulting in *jargon aphasia*. This jargon is usually meaningless, as for example in this response of a patient asked what a pen was used for: "This is a tape of brouse to make buke deproed in the auria" (Kertesz & Benson, 1970). The form and often the intonation makes such speech sound like sensible English, even though it is nonsense. Another characteristic of a person with Wernicke's aphasia is that they are unconcerned by the mistakes that they are making.

MOVEMENT AND SENSATION

In the 1940s, neurosurgeon Wilder Penfield and his neurologist colleague Herbert Jasper developed a technique of brain mapping to help reduce side effects caused by surgery to treat epilepsy, they stimulated motor and somatosensory cortices of the brain with small electrical currents to activate discrete brain regions. They found that stimulation of one hemisphere's motor cortex produces muscle contraction on the opposite side of the body. Furthermore, the functional map of the motor and sensory cortices is fairly consistent from person to person; Penfield and Jasper's famous pictures of the motor and sensory homunculi were the result.

LATERALIZATION OF LANGUAGE

The area of the brain that is responsible for both spoken and written language is the Wernicke's Area in the left hemisphere. Analysis of written words is performed in the temporoparietal cortex as well as in the anterior inferior frontal cortex. On the other hand, speech production is initiated in the Broca's Area. This area is assisted by the premotor area in selecting and sequencing speech sounds. The arcuate fasciculus makes it possible for the language information in the Wernicke 's area to be transmitted to the Broca's Area to produce speech.

- **Lateralization:** Human split-brain studies have helped develop knowledge about language and lateralization. In split-brain studies, the cutting of the corpus callosum (a group of nerve fibers connecting the two brain hemispheres) is cut. These studies have proven that the left and the right brain hemispheres have specific language functions.
- **Left Hemisphere:** Naming objects is one of the language-related functions of the left hemisphere. Objects placed in the right visual field are easily recognized by both normal people and split-brain subjects. However, split-brain subjects cannot identify objects located in the left visual field unlike normal subjects. This proves that the only known language function of the left-hemisphere is to name identified objects. Logic, critical thinking and reasoning are also functions that are dominantly processed in the left hemisphere.
- **Right Hemisphere:** Verbal identification of objects presented in the left visual field cannot be identified by people who have undergone the split-brain surgery. However, they can identify these objects by means of their sense o touch. Some words can also be comprehended through the right hemisphere. The split-brain studies show that the figurative sides and context of language

Figure 6. Right and left hemisphere

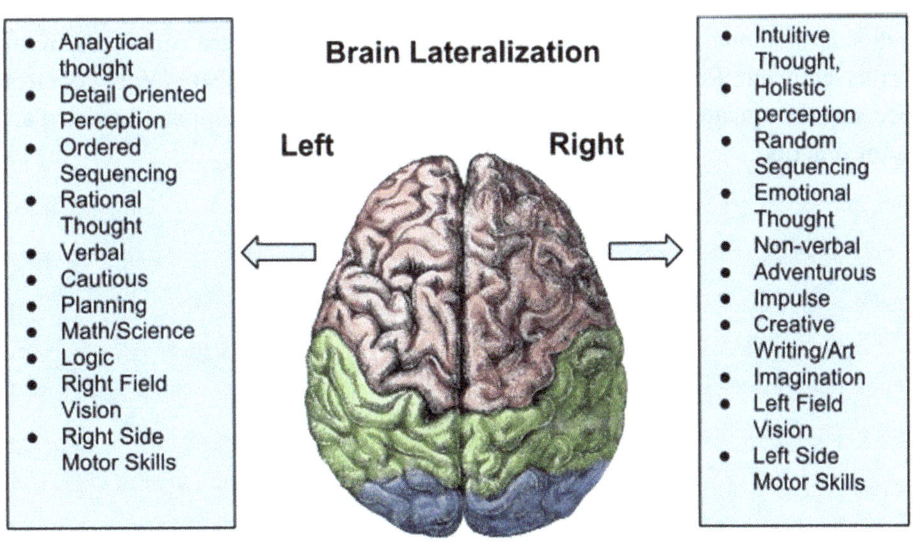

are understood via the right hemisphere. In addition, the emotional expression of language is processed in the right hemisphere. Also, music stimulates the right hemisphere more than spoken words do. Hemispheric lateralization is important in both the parallel processing and sharing of information. While the left hemisphere is mostly concentrated on the interpretation of information through logic and analysis, it coordinates with the right hemisphere which is focused on the interpretation of experience in its totality via synthesis.

- **Handedness:** In terms of handedness, most people who are right-handed reveal left hemisphere language dominance. This means that they tend to be better in logical and critical thinking than in creative and expressive thinking. Most left-handed people do not show right-handed language dominance, as common belief tells us. Rather, 70% of left-handed people still demonstrate left hemisphere language dominance (Fig 6)

DIFFERENCE BETWEEN LOCALISATION AND LATERALISATION IN THE BRAIN

Localisation of function in the brain is the theory that different areas of the brain control specific physical and behavioural functions in the body. The brain is divided into four lobes: the temporal lobe, occipital lobe, parietal lobe and frontal lobe. As an example, there is evidence to suggest that the occipital lobe is implicated in visual functions (seeing) - whereas, there is evidence that the parietal lobe is implicated in auditory functions (speaking and hearing). In contrast, lateralisation is the theory that different hemispheres of the brain control specific physical and behavioural functions in the body. The brain is split in half down the middle, and each side is known as a hemisphere - the two hemispheres are connected by the corpus callosum. Research has suggested that the left hemisphere largely controls speech and language, whereas the right hemisphere is more dominant in visual and motor functions.

APPROACHES AND METHODOLOGIES OF CLINICAL AND COGNITIVE NEUROPSYCHOLOGISTS

In the early 20th Century, neurologists were studying brain damaged patients for treatment purposes. Today, cognitive neuropsychologists have a number of goals depending on the type of work they are doing. Clinical neuropsychologists work with patients who have suffered brain damage and are interested in trying to get a good overall profile of the patients problems and strengths with a view to providing appropriate support.

Broadly, there are four main goals of clinical cognitive neuropsychologists:

- Lesion localization
- Assessment of a patients deficit
- Building models of normal cognition
- Localization of different cognitive functions within the brain

APPROACHES

Experimental neuropsychology is an approach that uses methods from experimental psychology to uncover the relationship between the nervous system and cognitive function. The majority of work involves studying healthy humans in a laboratory setting, although a minority of researchers may conduct animal experiments. Human work in this area often takes advantage of specific features of our nervous system (for example that visual information presented to a specific visual field is preferentially processed by the cortical hemisphere on the opposite side) to make links between neuroanatomy and psychological function.

Clinical neuropsychology is the application of neuropsychological knowledge to the assessment (neuropsychological test and neuropsychological assessment see the chapter-13), management, and rehabilitation typically specialized by clinical neuropsychologists, these people who have suffered illness or injury (particularly to the brain) which has caused neurocognitive problems. In particular they bring a psychological viewpoint to treatment, to understand how such illness and injury may affect and be affected by psychological factors. Clinical neuropsychologist's can offer an opinion as to whether a person is demonstrating difficulties due to brain pathology or as a consequence of an emotional or another (potentially) reversible cause or both. Clinical neuropsychologists often work in hospital settings in an interdisciplinary medical team; others work in private practice and may provide expert input into medico-legal proceedings.

Cognitive neuropsychology is a relatively new development and has emerged as a distillation of the complementary approaches of both experimental and clinical neuropsychology. It seeks to understand the mind and brain by studying people who have suffered brain injury or neurological illness. One model of neuropsychological functioning is known as functional localization. This is based on the principle that if a specific cognitive problem can be found after an injury to a specific area of the brain, it is possible that this part of the brain is in some way involved. However, there may be reason to believe that the link between mental functions and neural regions is not so simple. An alternative model of the link between mind and brain, such as parallel processing, may have more explanatory power for the workings and dysfunction of the human brain. Yet another approach investigates how the pattern of errors produced by brain-damaged individuals can constrain our understanding of mental representations and processes without reference to the underlying neural structure.

METHODS

- **Standardized Neuropsychological Tests:** These tasks have been designed so the performance on the task can be linked to specific neurocognitive processes. These tests are typically standardized, meaning that they have been administered to a specific group (or groups) of individuals before being used in individual clinical cases. The data resulting from standardization are known as normative data. After these data have been collected and analyzed, they are used as the comparative standard against which individual performances can be compared. Examples of neuropsychological tests include: the Wechsler Memory Scale (WMS), the Wechsler Adult Intelligence Scale (WAIS), Boston Naming Test, the Wisconsin Card Sorting Test (WCST), the Benton Visual Retention Test (BVRT), and the Controlled Oral Word Association (COWA). Barcelona Neuropsychological Test (BNT), Cambridge Neuropsychological Test Automated Battery (CANTAB),COGNISTAT (The Neurobehavioral Cognitive Status Examination), Cognitive Assessment Screening Instrument (CASI), Cognitive Function Scanner (CFS), Dean-Woodcock Neuropsychology Assessment System (DWNAS), General Practitioner Assessment Of Cognition (GPCOG), Hooper Visual Organization Test, Luria-Nebraska Neuropsychological battery, MICROCOG, Mini mental state examination (MMSE), NEPSY, Repeatable Battery for the Assessment of Neuropsychological Status, CDR Computerized Assessment System. In India widely used tests are PGI Battery of Brain Dysfunction, By Dwarka Pershad and S.K.Verma.(1990 & 2007) AIIMS Comprehensive Neuropsychological

Battery in Hindi [Adult Form] by Surya Gupta et al (2000) and NIMHANS Neuropsychology Battery - 2004 by Rao, Shobini L, et al (2004). Based on these test findings results usually correlated with routine Brain scans (see the chapter-12).

REFERENCES

Beaumont, J. G. (2008). *Introduction to Neuropsychology* (2nd ed.). The Guilford Press.

Braun, C. M., Delisle, J., Guimond, A., & Daigneault, R. (2009). Post unilateral lesion response biases modulate memory: Crossed double dissociation of hemispheric specialisations. *Laterality, 14*(2), 12264. doi:10.1080/13576500802328613 PMID:18991140

Dehaene, S., Spelke, E., Pinel, P., Stanescu, R., & Tsivkin, S. (1999). Sources of mathematical thinking: Behavioral and brain-imaging evidence. *Science, 284*(5416), 9704. doi:10.1126cience.284.5416.970

Devinsky, O. (2009). Delusional misidentifications and duplications: Right brain lesions, left braindelusions. *Neurology, 72*(1), 807. doi:10.1212/01.wnl.0000338625.47892.74 PMID:19122035

Gibson, W. C. (1962). Pioneers of localization of function in the brain. *Journal of the American Medical Association, 180*, 944–951. PMID:13898444

Halpern, M. E., Gunturkun, O., Hopkins, W. D., & Rogers, L. J. (2005). Lateralization of the Vertebrate Brain: Taking the Side of Model Systems. *The Journal of Neuroscience: The Official Journal of the Society for Neuroscience, 25*(45), 10351–10357. doi:10.1523/JNEUROSCI.3439-05.2005 PMID:16280571

Hecht, D. (2010). Depression and the hyperactive right-hemisphere. *Neuroscience Research, 68*(2), 77–87. doi:10.1016/j.neures.2010.06.013 PMID:20603163

John, J. P., & Pinel, P. (2011). *Biopsychology* (8th ed.). Academic Press.

Kandel, E., Schwartz, J., & Jessel, T. (2000). *Principles of Neural Science* (4th ed.). New York: McGraw–Hill.

Kertesz, A., & Benson, D. F. (1970). Neologistic jargon: A clinicopathological study. *Cortex, 6*(4), 362–386. doi:10.1016/S0010-9452(70)80002-8 PMID:5514511

McGilchrist, I., & Shankar, V. (2019). One Head, Two Brains: How The Brain's Hemispheres Shape The World We See (Audio podcast with transcript). *NPR Hidden Brain.*

Pershad, D., & Verma, S. K. (2007). Hand Book of PGI Battery of Brain Dysfunction (PGIBBD) (2nd ed.). National Psychological Corporation.

Rao, S.L., Subbakrishna, D.K., & Gopukumar, K. (2004). *NIMHANS Neuropsychological Battery-Manual.* National Institute of Mental Health and Neurosciences.

Richard, A. G., & Brian, C. (2012). Following Teaching of Psychology Into Middle Age: Changes, Contributors, and Contents. *Teaching of Psychology*, *40*(1), 6–14.

Ries, S. K., Dronkers, N. F., & Knight, R. T. (2016). Choosing words: Left hemisphere, right hemisphere, or both? Perspective on the lateralization of word retrieval. *Annals of the New York Academy of Sciences*, *1369*(1), 111–131. doi:10.1111/nyas.12993 PMID:26766393

Taylor, I., & Taylor, M. M. (1990). *Psycholinguistics: Learning and using Language*. Pearson.

KEY TERMS AND DEFINITIONS

Brain: A brain is an organ that serves as the center of the nervous system in all vertebrate and most invertebrate animals. It is located in the head, usually close to the sensory organs for senses such as vision. It is the most complex organ in a vertebrate's body. In a human, the cerebral cortex contains approximately 14–16 billion neurons, and the estimated number of neurons in the cerebellum is 55–70 billion. Each neuron is connected by synapses to several thousand other neurons. These neurons communicate with one another by means of long protoplasmic fibers called axons, which carry trains of signal pulses called action potentials to distant parts of the brain or body targeting specific recipient cells.

Cerebral Cortex: The cerebral cortex is the thin layer of the brain that covers the outer portion (1.5mm to 5mm) of the cerebrum. It is covered by the meninges and often referred to as gray matter. The cortex is gray because nerves in this area lack the insulation that makes most other parts of the brain appear to be white

Cognitive Psychology: *Cognitive psychology* is the scientific study of mental processes such as attention, language use, memory, perception, problem solving, creativity, and thinking.

Lateralization: The lateralization of brain function is the tendency for some neural functions or cognitive processes to be specialized to one side of the brain or the other. The medial longitudinal fissure separates the human brain into two distinct cerebral hemispheres, connected by the corpus callosum.

Localization: Localization of function is the idea that certain functions (e.g., language, memory, etc.) have certain locations or areas within the brain. This idea has been supported by recent neuroimaging studies, but was also examined much earlier, typically using case studies.

Neuropsychology: Is a branch of psychology that is concerned with how the brain and the rest of the nervous system influence a person's cognition and behaviors. More importantly, professionals in this branch of psychology often focus on how injuries or illnesses of the brain affect cognitive functions and behaviors.

Chapter 8
Temporal Lobe Syndromes:
An Overview

Arpit Parmar
All India Institute of Medical Sciences (AIIMS), India

G. S. Kaloiya
All India Institute of Medical Sciences (AIIMS), India

Harsimarpreet Kaur
https://orcid.org/0000-0003-2757-1754
Tata Institute of Social Sciences, India

ABSTRACT

Temporal lobes are one of the four major lobes of the cerebral cortex and perform a complex array of interrelated functions. They play an important role in various day-to-day functioning. The common pathologies leading to isolated temporal lobe dysfunction are infarction (of the middle cerebral artery), hemorrhage, seizures, tumors, encephalitis, and traumatic brain injury. Temporal lobe syndromes include a wide array of various neurological (Kluver-Bucy syndrome, Geschwind Gastaut syndrome, etc.), elementary (e.g., vertiginous syndromes, hallucinations, etc.), neuropsychiatric (e.g., anxiety, agitation, aggression, etc.), and cognitive (e.g., Korsakoff amnesia, cortical deafness, etc.) disorders. The presentation depends on a multitude of factors including involvement of dominant or non-dominant lobe. Left temporal lobe involvement usually leads to various forms of aphasia while right side involvement leads to more covert and varied syndromes. In this chapter, the authors discuss the anatomy of the temporal lobe, its functional aspects, and various syndromes of temporal lobe dysfunction.

DOI: 10.4018/978-1-7998-2860-0.ch008

CHAPTER OUTLINE

1. Introduction
2. Anatomy of temporal lobe
3. Functional anatomy of temporal lobe
4. Connections of the temporal lobe
5. Temporal lobe syndromes
    a. Kluver-Bucy syndrome
    b. Geschwind-Gastaut syndrome
    c. Wernicke's aphasia
    d. Cortical deafness
    e. Delusional misidentification
    f. Auditory paracusias
    g. Auditory agnosia
    h. Amusia
    i. Prosopagnosia
6. Neuropsychology of the temporal lobe: Assessment and its implications
7. Conclusion

INTRODUCTION

The brain's cerebral cortex is the outermost part of the brain which contains two cerebral hemispheres connected by the corpus callosum. Traditionally, each of these cerebral hemispheres is divided into four lobes: Frontal, parietal, temporal and occipital. The frontal lobe is located in the most anterior part of the brain and extends to the central sulcus. The frontal lobe is involved in language, emotion, motor control, and reasoning. The frontal lobe contains the motor cortex, prefrontal cortex, and Broca's area. The parietal lobe is located behind the frontal lobe and contains the somatosensory cortex, essential in sensory processing such as touch, pain, and temperature. The occipital lobe is located in the most posterior part of the brain and contains the primary visual cortex. The temporal lobe is located on the side of the head and is important in multiple aspects of brain functions such as hearing, memory, emotions, language, learning, time perception, special senses, and motivation. Damage to the temporal lobe can be caused by a multitude of cerebral pathologies including seizures, infarction, hemorrhage, trauma, encephalitis, etc. Thus, the temporal lobe damage can lead to impairment in one or several areas of brain functioning depending on the nature, extent, site, and laterality of the lesion. Here we describe the various types of temporal lobe syndromes and relevant functional neuroanatomy of the temporal lobe. We also provide a brief overview of the specific temporal lobe syndromes and the neuropsychological assessment in such patients.

ANATOMY OF TEMPORAL LOBE

The temporal lobe is the second-largest lobe and accounts for more than 20% of the total neocortex volume. The lobe extends to the Sylvian fissure superiorly and to an imaginary line posteriorly. The lateral parietal-temporal line separates the temporal lobe from the parietal lobe superiorly and occipital lobe inferiorly. Its anterior and inferior boundaries are formed by the middle cranial fossa. The temporal lobe is divided into two sections: neocortex and mesial temporal lobe. The mesial/medial temporal lobe contains areas such as the hippocampus, parahippocampal gyrus, amygdala, etc. and is called the limbic lobe as well. The temporal lobe is divided by four sulci into five gyri. The five gyri are superior temporal gyrus, middle temporal gyrus, inferior temporal gyrus, fusiform gyrus, and parahippocampal gyrus. The temporal lobe receives blood supply from both the internal carotid artery (anterior choroidal artery which supplies the anterior part of the parahippocampal gyrus, the amygdala and the uncus and middle cerebral artery which supplies the superior and inferior temporal gyri) and the vertebrobasilar artery (supplies the inferior surface). The temporal lobes are tied to other lobes intimately through various association tracts. The largest of these association tracts include the occipitotemporal and uncinate fasciculus.

FUNCTIONAL ANATOMY OF TEMPORAL LOBE

The temporal lobe is located just beneath the lateral fissure and crosses both fissures of the brain. It plays a vital role in many. The temporal lobe is a vital structure that helps in processing sensory input, including pain and auditory stimuli. It also helps in understanding the language, retaining the visual memories. It is also involved in both processing and remembering emotions. It's a structure where the sound is processed and where auditory language and speech comprehension systems are located. The auditory cortex is found within the Sylvian fissure on the upper banks of the temporal lobe. Wernicke's area is located to the posterior of the auditory complex which is an area for speech comprehension. The temporal lobe's middle regions are supposed to contain conceptual representations for semantic information. The posterior and inferior areas of the temporal lobe are finetuned for representing the visual objects. The medial temporal lobe (MTL) is a part of the temporal lobe which has distinct functions and anatomy and hence it is typically referred to as a separate structure. The MTL plays a crucial role in declarative memory and serves as a temporary deposit of information while the final storage of information is done at a neocortex level. Thus, the interaction between the MTL and neocortex ensures the encoding-storage-retrieval processes for declarative memory. The MTL is also home to a region called the limbic region. The word "limbus" means boundary, however, there is a great deal of debate about the structures forming the limbic region. The entire complex of hippocampi, limbic complex, and amygdala is sometimes called the limbic system. The structures and interacting areas of the limbic system are involved in memory, emotion, motivation, and learning. The MTL also contains hippocampi, a structure that has a role in memory function of the brain. The hippocampus is fundamental in the formation of declarative memory, in both episodic and semantic aspects. Thus, the temporal lobes play an important role in virtually every brain function and damage to this area can have many consequences depending on the region of the temporal lobe involved.

CONNECTIONS OF THE TEMPORAL LOBE

The temporal lobes are rich in internal connections, afferent projections from the sensory systems, and efferent projects to the parietal and frontal association regions, limbic system and the basal ganglia.

The five distinct types of temporal-cortical connections are given below:

1. **A hierarchical sensory pathway:** The hierarchical progression of connections emanates from the primary and secondary auditory and visual areas, ending in the temporal pole. The visual projections form the ventral stream of visual processing, whereas the auditor projections form a parallel ventral stream processing.
2. **A dorsal auditory pathway:** Traveling from the auditory areas to the posterior parietal cortex, this pathway is analogous to the dorsal visual pathway and thus concerned with directing movements with respect to auditory information.
3. **A polyvagal pathway:** This pathway is a series of parallel projections from the visual and auditory association areas into the polyvagal regions of the superior temporal sulcus.
4. **A medial temporal projection:** The projection from the auditory and visual association areas into the medial temporal, or limbic, regions goes to the hippocampal formation or the amygdala or both. The hippocampal projection is a major one, forming the performant pathway. A disturbance to this projection results in a major hippocampal activity.

5. **A frontal-lobe projection:** This series of parallel projections reach from the association areas to the frontal lobe.

TEMPORAL LOBE SYNDROMES

The common pathologies which affect the temporal lobes in insolation include infarction (of middle cerebral artery inferior division), epilepsy, tumors, traumatic brain injury, intracerebral hemorrhage, and encephalitis. Apart from the aphasia which is typical of involvement of the left temporal lobe, the right temporal lobe syndromes are more covert and enigmatic. Various kinds of neurological deficits occur with unilateral or bilateral lesions of the temporal lobes. Unilateral lesions in the dominant hemisphere lead to deficits such as acalculia, agraphia, alexia, nominal dysphasia, Wernicke's dysphasia, and contralateral homonymous superior quadrantanopia among many others. On the other hand, unilateral lesions in the non-dominant hemisphere cause deficits such as contralateral homonymous superior quadrantanopia and prosopagnosia. Irrigative lesions in either lobe can lead to visual hallucinations, seizures (complex-partial variety) and various memory disturbances. Different forms of syndromes of the temporal lobe are summarized in table 1.

Kluver-Bucy syndrome

Kluver-Bucy syndrome (KBS) is a rare behavioral impairment that is associated with lesions involving bilateral involvement of anterior temporal lobes of the brain. However, cases with unilateral involvement have also been reported. The presentation includes symptoms of hyperorality, hyper/ hyposexuality, visual agnosia (inability to visually recognize objects), amnesia, placidity, loss of normal fear and anger responses, hyper metamorphosis, distractibility, seizures, and dementia. KBS may be permanent or transient. The KBS is caused by the trauma or other pathologies involving temporal lobe including tumors, degenerative brain diseases and infections (most commonly herpes simplex encephalitis). The treatment is primarily symptomatic and supportive. However, there is no cure for this condition. The disorder is not life-threatening but it can be difficult to manage at times.

Geschwind-Gastaut Syndrome

The Geschwind-Gastaut syndrome (GGS) is a disorder long been described in the context of temporal lobe epilepsy (TLE). TLE is associated with chronic, mild changes in personality during the interictal period (i.e. between the seizures). More recently it has also been demonstrated to occur even in isolation (after infarct and hemorrhage involving temporal lobe). Right temporal lobe pathologies have been associated with this syndrome more commonly. The cardinal features of this syndrome are vicious personality, metaphysical preoccupation, and altered physiological drives. The viscous personality in this syndrome is characterized by circumstantiality, excessive or overinclusive detailed narration of information, hypergraphia, undue prolongation of interaction, etc. Hypergraphia is a tendency for extensive/compulsive writing or drawing seen in patients with TLE. Metaphysical preoccupation may present with excessive and intense intellectual pursuits in relation to religion, morality, and philosophy. Some TLE auras may also involve ecstatic experiences. Abnormalities in sexuality have also been reported with hyposexuality being more common than hypersexuality.

Wernicke's Aphasia

Wernicke's aphasia is also known as receptive aphasia. Traditionally, aphasia is divided into two types: Broca's aphasia (expressive aphasia) and Wernicke's aphasia (receptive aphasia). An ischemic stroke affecting the posterior part of the dominant temporal lobe (Brodmann's area 22) is the most common cause of this condition. Wernicke's aphasia is characterized by impairment in language comprehension. However, the rate, rhythm, and grammar of the speech might be normal in this condition. The patients with this condition do not recognize the errors because of the impaired comprehension of their speech. Paraphasic errors are common which are generally of two types: Phonemic and semantic. More severe cases may lead to neologisms. Reading is also impaired as it involves comprehension of the written words. Associated symptoms may include symptoms of visual field deficits, acalculia, and agraphia. Depression and emotional outbursts are less common as compared to the Broca's aphasia. Traditionally motor deficits are absent in pure receptive aphasia. The Boston Diagnostic Aphasia Examination is one of the most commonly used assessment tools for language deficits. Speech and language therapy is the mainstay of the treatment as there is no standard treatment for this condition. Remediation efforts are challenging owing to the comprehension difficulties.

Cortical Deafness

The cortical or central deafness is a rare variety of hearing impairment described first by Wernicke and Friedlander who reported a case of a patient with bilateral temporal lobe lesion presenting with no awareness of sounds with no apparent damage to the hearing organ. The clinical presentation of this condition is similar to cochlear-type profound deafness. They show no consistent response to sounds and show raised thresholds on audiometry. Peripheral detection of sounds is completely

Table 1. Temporal lobe syndromes: Overview of deficits based on the hemisphere involvement

Right temporal lobe	Left temporal lobe	Bilateral or right or left
• Left upper quadrantanopia **• Amnesia (visuospatial variety)** **• Amusia** **• Verbal and non-verbal auditory agnosia** **• Prosopagnosia** **• Theory of mind impairment** **• Delusional misidentification syndromes**	• Right upper quadrantanopia • Wernicke's aphasia • Transcortical sensory aphasia • Anomic aphasia • Lexical amusia • Verbal amnesia • Synesthesia • Visual agnosia	• Kluver-Bucy syndrome • Korsakoff amnesia • Auditory agnosia • Auditory hallucinations/ paracusias • Cortical deafness • Time perception abnormalities • Geschwind-Gastaut syndrome • Taste abnormalities due to seizures or insula lesions • Olfactory hallucinations • Vertiginous syndromes • Neuropsychiatric manifestations such as paranoia, aggression, anxiety, agitation

or partially intact in these patients. The most common anatomical cause for this condition is vascular lesions involving primary, secondary, or association cortices in Heschl's gyrus and the planum temporale bilaterally. This may also occur after lesions resulting in the disconnection between sensory and perceptual processing. The diagnosis requires demonstration of normal brainstem auditory evoked responses along with impaired cortical evoked potentials (also known as brainstem auditory evoked potentials). The patients recover to a variable degree depending on the size, site, and type of the lesion.

Delusional Misidentification

Delusional misidentification syndromes refer to a false and fixed belief in doubles and duplicates. The syndrome includes Capgras syndrome and Fregoli syndrome along with their variants and reduplicative paramnesia. The syndromes are rooted in a disruption between the connections between cortical association areas and limbic structures located in the temporal lobe (that relates sensory information with the emotional valance). Thus, there appears to have a sensory-affective dissonance in which the patient has perceived the stimulus but lacks the emotional significance and relevance to self. These syndromes are often comorbid with anosognosia, other delusions, depersonalization, and derealisation, as well as environmental disorientation.

Auditory Paracusias

These may include hallucinations as well as illusions. Both simple, as well as complex auditory hallucinations, may occur. Peduncular hallucinosis is a possible differential for this condition which usually presents with visual hallucinations while auditory hallucinations are less common.

Auditory Agnosia

Like all other agnosias, auditory agnosia means there is a failure of recognition and not perception. Patients with this condition are not able to interpret the meanings of speech and peripheral sounds. Thus, a patient with auditory agnosia may be able to hear the sound of an alarm clock but he/she does not recognize the nature of the sound and experience it simply as noise. Unlike patients with cortical/central deafness, these patients may have a normal hearing as measured by the pure tone audiometry. This is usually caused by a lesion involving bilateral temporal lobes. The anatomical pattern of damage is similar in auditory agnosia, cortical/central deafness, and pure word deafness.

Amusia

Amusia is a condition in which there is a loss of musical ability secondary to a brain disease. The subtypes of amusia include receptive, expressive, and lexical amusia. Receptive amnesia is defined as the loss of ability to discriminate between the melodies while amnesic amusia is defined as the loss of ability to identify and discriminate between the familiar melodies. Expressive amusias include oral-expressive amusia, musical apraxia, and musical agraphia. Oral-expressive amusia is a loss of the ability to sing, whistle or hum a tone. Musical apraxia is the lost ability to play a musical instrument while musical agraphia is described as a loss of music writing ability. Another form of amusia, musical anhedonia, is defined as a selective emotional impairment in listening to the music. Though the language function and melody recognition are preserved, these patients are unable to sing, hum, or whistle. Expressive amusia usually occurs secondary to right temporal lobe lesions and there are numerous case reports while receptive amusia may also occur after right-sided lesions.

Prosopagnosia

Prosopagnosia is selective visual agnosia. The patient suffers from an inability to recognize the identity of faces. The impairment affects both faces well known to them as well as those recently encountered. The subjects realize that a face is a face and not a tree or a car, but cannot say whom that face belongs to and can't identify the person. They rely on other features to identify a person such as hair, gait, voice, etc. The most commonly used test for the familiarity with the recently seen faces is the Cambridge face Memory Test while the Cambridge Face Perception Test is used for the assessment of deficits in face perception, i.e. the ability to perceive differences between faces. Most treatment attempts focus on enhancing the coping strategies to improve face recognition. The training efficacy depends on multiple factors including age, age of onset, laterality, type of lesion, and location.

NEUROPSYCHOLOGY OF THE TEMPORAL LOBE: ASSESSMENT AND ITS IMPLICATIONS

Kolb and Wishaw (1990) have identified eight principle symptoms of temporal lobe damage: 1) disturbance of auditory sensation and perception, 2) disturbance of selective attention of auditory and visual input, 3) disorders of visual perception, 4) impaired organization and categorization of verbal material, 5) disturbance of language comprehension, 6) impaired long-term memory, 7) altered personality and affective behavior, 8) altered sexual behavior. Table 2 provides an overview of the tests which can be utilized to assess the neuropsychological impairments caused by the damage to these functions.

Luria and modern scientists used studies of lesions and pathologies of the cerebral organs as the basis for the development of neuropsychology. However, nowadays, due to the emergence of new imaging techniques that also permit the study of the cerebral lobes, which corroborates studied data on the function of the temporal lobe in normal speech. Furthermore, modern surgical techniques allow the study of more patients that survive after these procedures due to lower post-surgical mortality rates compared to those of surgical procedures utilizing older techniques. Modern studies involve epilepsy, as a temporal lobe syndrome with neuropsychological concerns. Neuropsychological data on lesions in the anterior part of the temporal lobe considering the hippocampus, obtained by functional magnetic resonance imaging analysis after surgical resection, revealed that, in cases of epilepsy linked to the activation of the left hippocampus, responsible for word encoding, they are linked to verbal memory deficits, whereas for the right temporal lobe, epilepsy results in the decline of visual memory.

In functional terms and neuropsychological aspects, the medial temporal lobes work to maintain novel information in the absence of a perceptual stimulus. Therefore, the lesions in these regions demonstrate that they are linked to short-term and long-term memory. On the other hand, there is a hypothesis that verbal memory is fractioned by the mesial temporal region, while verbal and non-verbal memory is not totally lateralized in the temporal lobes. Brain lesions in epilepsy are associated with neuropsychological tests used to assess the possible risks of postsurgical cognitive decline and enable the evaluation of patients before and after the surgery in relation to cognition and memory. A clinical assessment could thus be used to verify the expectations of the patient's post-surgery. Recent neuropsychological tests on individuals with epilepsy take into account processes such as intellectual abilities, attention, visuospatial skills, language, executive function, and memory. The analysis of the temporal lobe in pre and post-surgical patients demonstrated a heterogeneous variation in relation to the neuropsychological tests used, for instance, Visual Naming, Logical Memory-Delayed, Logical Memory-Immediate, and Learning Test. In terms of the reorganization of functions in cases of cortical lesions, in patients with medial temporal lobe impairment, autobiographical memory became independent of this lobe. In patients with epilepsy, evaluations employing neuropsychological methods show that the decline in cognition is greater in the left temporal lobe than the right, in association with depression, psychiatric comorbidity undertreated in cases of epilepsy. The right and left temporal lobes seem to be associated with the perception of scary music because the lesions in these regions diminish or deplete the patients' capacity to recognize aggressive songs. Regardless of Luria's studies on the effects of sounds and music on the temporal lobe, the different aspects of the kind of music language were not verified, but he did cite the theory of both hemispheres working together in many processes. Pathological problems are mainly associated with the cerebral networks discerned through neuropsychological tests, particularly those used for memory in the temporal lobe, with demonstrations on the reorganization of language involving increased pre-frontal activation. Resection of part of the left temporal lobe generates memory loss, specifically verbal memory which increases with aging and causes impairment of language in part of, or for the remainder of, the patient's lifetime, an effect also found in lesions of the uncinate fasciculus, related to memory loss. Loss of semantic memory is a result of the effects of atrophy of the anterior temporal lobe, mainly the left side, as language impairment is associated with the region of the collateral sulcus and the anterior fusiform gyrus, part of the semantic system, not with the ventromedial frontal cortex. However, it seems to be a more extensive area encompassing a more extensive network. Memory problems with lesions in the temporal lobe, and the association with superior activities of the brain in the pre-frontal area, were extensively cited by Luria, albeit not necessarily with all of the frontal cortex.

Table 2. Neuropsychological assessment for temporal lobe syndromes

Domains	International Neuropsychological Tests	Available Indian Tests/ Batteries
Intelligence	Vocabulary, similarities, Block Design, Digit Symbol	Verbal Adult Intelligence Scale, MISIC for children
Declarative memory	Logical memory 1 (immediate) and II (delayed) Faces I (immediate) and II (delayed) Visual Reproduction I (immediate) and II (delayed) RAVLT Total and delayed recall	WMS- 3 Indian Edition AVLT
Language	Boston Naming Test (BNT) Semantic Category (Animals) Western Aphasia Battery (WAB) Boston Aphasia Diagnostic Examination (BDAE) Minnesota Test for the Differential Diagnosis of Aphasia (MTDDA)	Western Aphasia Battery –Hindi Adaptation, Kannada Adaptation Indian Aphasia Battery Tests for language from the NIMHANS battery: Animal Naming Test, Controlled Word Association Test, Token Test
Nonverbal memory and Visuospatial functioning	Rey's Complex Figure Test	Complex Figure Test

Left TLE lesions, a regularly occurring even in epilepsy disorder and the most common type of medically intractable partial epilepsy, on top of hippocampal sclerosis, leads to verbal memory impairment in patients. Nevertheless, fMRI activation of the ipsilateral hippocampus in left TLE patients correlates with better memory when compared with contralateral hippocampus activation. In fact, there are data suggesting that hippocampal damage is the primary cause of deficiencies in non-verbal memory. Thus, both medial temporal lobes seem to be responsible for linking different cortical regions of the representation of memory. In fact, lesions in these regions cause amnesia. These results can also be found in amygdalohippocampectomy of patients with temporal lobe epilepsy. However, it is important to bear in mind that the lateralization of these functions was found in adolescents and adults, as well as pre-adolescents; and the hypothesis of the lateralization of the anterior temporal lobe was verified in relation to facial recognition, naming of the same entity and memory deficiencies.

Studies on the effects of temporal lobe lesions and hippocampus atrophy caused by Alzheimer's show a strong correlation between mini-mental and memory tests, indicating an important means of diagnosing Alzheimer's. For providing a qualitative rating of medial temporal lobe atrophy, it is a fairly good diagnostic tool. In conjunction, the amygdala and hippocampus are associated with verbal intelligence deficiencies and increased anxiety and depression. Nevertheless, mesial temporal lobe epilepsy and hippocampal sclerosis cause intellectual impairment as a non-specific symptom. Notwithstanding, there are reports that surgical treatment of temporal lobe epilepsy is ineffective, according to neuropsychological tests for verbal and visual memory, at least for small samples. Interestingly, bilateral disturbances in the temporal lobe could be associated with hypersexuality.

CONCLUSION

The temporal lobe plays an integral role in the neuropsychological functioning of the individual. Because the temporal lobe is involved in so many bodily functions, damage to the temporal lobe can be difficult to manage. The effects of temporal lobe damage are in plenty and can be clinically significant. Cognitive assessment can assist in diagnosis (i.e., localization of dysfunction) and in prediction (i.e., how cognition may change following surgery) in individuals being considered for temporal lobe syndromes. The current practice includes behavioral testing as well as mapping function via stimulation, inactivation, and functional imaging. It can also be helpful in providing a clear picture of the management and rehabilitation of temporal syndromes.

REFERENCES

Akanuma, N., Alarcón, G., Lum, F., Kissani, N., Koutroumanidis, M., Adachi, N., ... Morris, R. G. (2003). Lateralising value of neuropsychological protocols for presurgical assessment of temporal lobe epilepsy. *Epilepsia*, *44*(3), 408–418. doi:10.1046/j.1528-1157.2003.24502.x PMID:12614397

Amadio, J. P., Beasley, K., Rothstein, B. D., Selman, W. R., Riley, J., & Emery, J. (2015). Intervening in the Brain: Ethics and Neurosurgery. *The Virtual Mentor*, *17*, 1.

Aversi-Ferreira, T. A., Tamaishi-Watanabe, B. H., Magri, M. P. D. F., & Aversi-Ferreira, R. A. (2019). Neuropsychology of the temporal lobe: Luria's and contemporary conceptions. *Dementia & Neuropsychologia*, *13*(3), 251–258. doi:10.1590/1980-57642018dn13-030001 PMID:31555397

Bakker, A., Albert, M. S., Krauss, G., Speck, C. L., & Gallagher, M. (2015). Response of the medial temporal lobe network in amnestic mild cognitive impairment to therapeutic intervention assessed by fMRI and memory task performance. *NeuroImage. Clinical*, *7*, 688–698. doi:10.1016/j.nicl.2015.02.009 PMID:25844322

Barnett, A. J., Park, M. T. M., Pipitone, J., Chakravarty, M. M., & McAndrews, M. P. (2015). Functional and structural correlates of memory in patients with mesial temporal lobe epilepsy. *Frontiers in Neurology*, *6*, 103. doi:10.3389/fneur.2015.00103 PMID:26029159

Bonelli, S. B., Powell, R. H., Yogarajah, M., Samson, R. S., Symms, M. R., Thompson, P. J., ... Duncan, J. S. (2010). Imaging memory in temporal lobe epilepsy: Predicting the effects of temporal lobe resection. *Brain*, *133*(4), 1186–1199. doi:10.1093/brain/awq006 PMID:20157009

Chaudhary, K., Kumaran, S. S., Chandra, S. P., Wadhawan, A. N., & Tripathi, M. (2014). Mapping of cognitive functions in chronic intractable epilepsy: Role of fMRI. *The Indian Journal of Radiology & Imaging*, *24*(1), 51. doi:10.4103/0971-3026.130694 PMID:24851005

Cole, M. (2017). *Selected Writings of AR Luria*. Routledge. doi:10.4324/9781315171913

Goldberg, E. (Ed.). (2019). *Contemporary neuropsychology and the legacy of Luria*. Psychology Press. doi:10.4324/9780203771662

Kellermann, T. S., Bonilha, L., Eskandari, R., Garcia-Ramos, C., Lin, J. J., & Hermann, B. P. (2016). Mapping the neuropsychological profile of temporal lobe epilepsy using cognitive network topology and graph theory. *Epilepsy & Behavior*, *63*, 9–16. doi:10.1016/j.yebeh.2016.07.030 PMID:27532489

McAndrews, M. P., & Cohn, M. (2012). Neuropsychology in temporal lobe epilepsy: Influences from cognitive neuroscience and functional neuroimaging. *Epilepsy Research and Treatment*, *2012*, 1–13. doi:10.1155/2012/925238 PMID:22957249

Moore, P. M., & Baker, G. A. (2002). The neuropsychological and emotional consequences of living with intractable temporal lobe epilepsy: Implications for clinical management. *Seizure*, *11*(4), 224–230. doi:10.1053eiz.2001.0668 PMID:12027568

Saling, M. M. (2009). Verbal memory in mesial temporal lobe epilepsy: Beyond material specificity. *Brain*, *132*(3), 570–582. doi:10.1093/brain/awp012 PMID:19251757

KEY TERMS AND DEFINITIONS

Kluver-Bucy Syndrome: Is a rare behavioral impairment characterized by inappropriate sexual behaviors and mouthing of objects. Other signs and symptoms, include a diminished ability to visually recognize objects, loss of normal fear and anger responses, memory loss, distractibility, seizures, and dementia.

Temporal Lobe: The *temporal lobe* is the region where sound is processed and, not surprisingly, it is also a region where auditory language and speech comprehension systems are located. The auditory cortex is located on the upper banks of the *temporal lobe* and within the Sylvian fissure.

Temporal Lobe Syndromes: The common pathologies which affect the temporal lobes in insolation include infarction (of middle cerebral artery inferior division), epilepsy, tumors, traumatic brain injury, intracerebral hemorrhage, and encephalitis.

Chapter 9
Parietal and Occipital Lobe Syndromes

Shahzadi Malhotra
Chacha Nehru Hospital, India

ABSTRACT

The human brain is the most important as well as the most complex organ in the human body. From previous chapters, we by now know that the cerebral cortex is the part of the brain that functions to make human beings unique. Distinctly human traits including higher thought, language, and human consciousness as well as the ability to think, reason, and imagine all originate in the cerebral cortex. The cerebral cortex is the outermost portion that can be divided into the four lobes of the brain. Each bump on the surface of the brain is known as a gyrus, while each groove is known as a sulcus. In this chapter, the authors discuss the parietal and occipital lobes of the cortex. It then highlights their functional issues.

CHAPTER OUTLINE

- Introduction
- Neuroanatomy of Parietal lobe
- Functioning of Parietal Lobe
- Parietal lobe dysfunction
- Neuroanatomy of Occipital lobe
- Functioning of Occipital Lobe
- Occipital lobe dysfunction

DOI: 10.4018/978-1-7998-2860-0.ch009

Learning Objectives

- To describe neuroanatomy of parietal lobe
- To describe the functioning of parietal lobe
- To explain the various signs and symptoms of parietal lobe dysfunction
- To describe neuroanatomy of occipital lobe
- To describe the functioning of occipital lobe
- To explain the various signs and symptoms of occipital lobe dysfunction

INTRODUCTION

Human brain is the most important as well as the most complex organs in the human body. From previous chapters, we by now know that the cerebral cortex is the part of the brain that functions to make human beings unique. Distinctly human traits including higher thought, language, and human consciousness as well as the ability to think, reason and imagine all originate in the cerebral cortex. The cerebral cortex is the outermost portion that can be divided into the four lobes of the brain. Each bump on the surface of the brain is known as a gyrus, while each groove is known as a sulcus. In this chapter, we discuss the parietal and occipital lobes of the cortex.

THE PARIETAL LOBE

The parietal lobe is the part of the cerebral cortex between the occipital and frontal lobes, and above the temporal lobe. It is called parietal lobe because of its proximity to the parietal bone, the wall of the skull (from the Latin *paries*, wall). The central sulcus lies anterior to the parietal lobe, inferiorly by the posterior end of the lateral sulcus, and posteriorly it is bounded by an imaginary borderline. The primary sensory, also known as somatosensory or somesthetic, cortex is found in the parietal lobe. It is also known as the *sensory strip*, but in general, the term *somatosensory* is used as it helps to associate it with its function and makes it easy to remenber its function- that bodily sensations, and not visual or auditory sensations, are processed here. This gyrus lies directly posterior to the central sulcus, or rolandic fissure. Sensory control of various body parts has been mapped on this sensory cortex. Somesthetic sensations (e.g., pain, temperature, touch) are sent to the sensory cortex from the opposite side of the body.

The parietal lobe is described by the three anatomical landmarks: The central sulcus divides the parietal lobe from the frontal lobe; the parieto-occipital sulcus divides the parietal and occipital lobes; the lateral sulcus is the most lateral boundary, dividing it from the temporal lobe. The medial longitudinal fissure separates the two hemispheres.

PARIETAL LOBE FUNCTION

The initial insights into the functions of parietal lobe came for the in depth analysis of the deficits of patients with lesions of the parietal lobe (Critchley, 1953). Subsequent understandings into the mechanisms of parietal function came from neuroanatomical and neurophysiological studies. Inside each hemisphere of parietal lobe, the somatosensory cortex serves the skin area on the contralateral surface of the body. Being divided into the left and right hemisphere, some definite structures play important roles in parietal lobe functioning. These areas with their respective functions are listed below:

Posterior Parietal Cortex coordinates movements and has important role in spatial reasoning which in turn plays a vital role in attention to the new stimulus.

Postcentral Gyrus area is the primary somatosensory cortex, which obtains sensory information and loads it on a sensory homunculus. This area is also designated to as Brodmann Area 3. *Inferior Parietal Lobule* helps in assessing facial expressions to judge emotions. It also assists in processing language, mathematical operations, and body language.

Superior Parietal Lobule assists in the determination of one's own spacial orientation as well as of different objects such as addresses and roads. It gains significant input by hand, that is, it supports the coordination of fine motor skills.

Primary Sensory Area where the impulse from the skin is obtained and interpreted, such as cold, warmth, touch, and pain. The parietal lobe is known for sensory perception and integration, including the taste, hearing, sight, touch, and smell. The parietal lobe is home to the brain's primary sensory area, an area where the brain interprets input from other areas of the body. Two gyri in the parietal lobe are associated with language. The first is the supramarginal gyrus, which curves around the posterior end of the lateral sylvian fissure. The second, the angular gyrus, lies directly posterior to the supramarginal gyrus. It curves around the end of a prominent sulcus in the temporal lobe, the superior temporal sulcus. Damage in the area of the angular gyrus in the dominant left hemisphere may cause word-finding problems (anomia), reading and writing deficits (alexia with agraphia), as well as left-right disorientation, finger agnosis (inability to identify the fingers), and difficulty with arithmetic (acalculia).

The postcentral gyrus (somatosensory cortex) is a *primary* cortical area, whereas the majority of the remaining parietal lobe cortex is composed of association cortex, and is largely concerned with somatosensory and visual association function. The parietal lobe is also known as the "association area of association areas" because of the multimodal processing that takes place in he parietal lobe. As a whole, the parietal lobes is considered to be the most lateralized in function of all the lobes, although the specialization is not complete. Language functions tend to be concentrated in the left parietal lobe around the gyri. In the nondominant hemisphere, the parietal lobe association cortex primarily processes spatial information and related selective attention. Deficits and damage in this part may result in difficulty attending to or a complete neglect of the contralateral side of space. Visuospatial and constructional deficits (such as in drawing, building small models, and so on) may be found in patients with right-hemisphere parietal lobe damage.

To summarize, parietal lobe has five major functions:

- The primary analysis of somatic sensation (touch, the position of the limbs, temperature);
- The analysis of space using all sensory modalities;
- The specification of spatial targets for the motor system;
- The generation of attention; and
- The analysis of visual motion

PARIETAL LOBE DAMAGE

The parietal lobe damage can lead to irreversible consequences. The prognosis depends on the site of damage, how severe the injury is and if treatment is possible or not. Thc carc may include occupational, speech and physical therapy. The following are the specific syndromes commonly found in patients having parietal lobe damage:

- **Constructional Apraxia**: Construction of two dimensional or three dimensional forms are affected in parictal loba lesion, which is morew prominent when the lesions are in the non -dominant partietal lobe. Constructional apraxia for two dimensional figures is present when the patient is unable to copy simple geometrical figures. Distortion rotation micrographia (the drawing is reduced in size) and macrographia (the drawing is enlarged in size) may be present,. Constructional apraxias for three dimensional figures are present when the patient is unable to construct three dimensional forms. The placement of either blocks or sticks even in the simole form of a square may not be possible. In the next stage, if forms are constructed they may be rotated. The form may also be constructed on the design itself, qualifying for the closing in phenomenon, the patient may or may not benefit from cues.

Usually constructional apraxia is associated with lesions of the right parictal lobe. If the patient substantially benefits from cues, it can be lateralized to the left partictal lobe. It can be assessed through complex figure and block design tests.

- **Holme's Syndrome:** Disturbances in the perception of absolute distance from oneself to an object is known as absolute localization. The distance between tow objects external to oneself is relative localization. Both these can be disturbed in right parietal lesions. These deficits are manifest when the patient is unable to grasp an object extended to him/her. The reaching movements are not accurate. It appears as if the patient has poor eyesight. Size constancy is lost. The patient perceives the bigger objects as nearer and the smaller objects as father.
- **Agnosia:** Lesions of the left angular gyrus or disonnection of this area from the occipital lobes results in an inability to recognize familial objects through the visual modality. The patient is able to recognize the object through another modality such a touch. This deficit is termed as visual object agnosia. Understanding of a complex scene or a picture may be impaired as in *simultagnosia*. The patient is able to describe the picture in parts, but is unable to sum up the totality of a scene. Recognitiuon of objects through touch is impaired in parietal lesions, known as tactile agnosia or *astereognosis*. Familiar objects placed in the hand are not recognised. The deficit may be unilateral wherein the lesion is in the contralateral parietal loba. If it is bilatera, the lesion is in the left p [artietal lobe. Finger agnosia is another condition wherein the patient isunable to name or identify the fingers which are touched. The patient is unable to identify his/ her own fingers and the fingers of the examiner. This is an inability to recognize body parts, is bilateral innature an associated with left partiertal lesions. *Anosognosia* is an inability to recognize a paralyzed limb as belinging to oneself. *Prosopagnosia*, the inability to recognize familiar faces visually is associated with right parietal lesions. The patient can recognize the same person through the clothes, vice or the silhouette. Recognition is absent when the faceis seen. The patient may even be unable to recognize his own face in the mirror.
- **Unilaterla Spatial Neglect:** An important sign of right parietal lesions is unilateral spatial neglect or hemi neglect or hemi inattention. Visual attention is not allocated voluntarily to one half of space. The patient is able to attend to it if the attention is drawn to this area verbally. Spontaneously attention is not allocated to the side contra lateral to the lesion. The neglect is usually present on the left half of space. The lesion is in the right parietal lobe. Visual attention is an important function mediated by the right parietal lobe. The right dorsal parietal lobe disengages attention in visual space, if this

disengagement does not occur, attention is fixated to one area and other areas are neglected. The attention disengagement results in the patient neglecting all aspects of the contra lateral space. The patient does not see objects on the neglected side and bumps into them. Dressing of one half is neglected. Drawings omit the left half of the figure. Touch is not felt when it is in the left half of the body. Hemi- inattention can occur in right occipital and right frontal lesions also.

- **Visual Memory Problems:** Right parietal lesions are associated with disturbances of visual memory. The localization is to right parietal area when the memory disturbance is for simple visual forms. Memory of places or locations i.e., topographical memory is disturbed in right parietal lesions. The patient is unable to locate familiar places on a map.
- **Route Finding Difficulty:** Independent of the memory deficit, patients with left parietal lesions get lost in familiar surroundings which are termed as *route finding difficulty*. It can occur even in their own lomes, the patient is able to verbally recall the reut but gets lost while traversing it. The deficit is hypothesized to arise from another parietal deficit which is left-right disorientation. Here the patient gets confused between the left and right sides. Identifying one's left or right side is difficult. The patients are confused when asked to identify the examiner's left or right side.
- **Apraxic Aphasic Alexia:** The left partictal lobe, in particular the angular gyrus is important for writing, reading and calculation, the semantic lexicon is situnted here. Lesions cause damage to the lexicon following which comprehension of oral and written language is affected. This condition is the *apraxic aphasic alexia*. If the arigular gyrus is disconnected from the visuo perceptual centers of the occipital lobe, only *alexia* is present. It is characterized by impaired reading. Disconnection for the left angular gyrus from the motor engrams situated in the inferior parietal lobule results in agraphia. Writing difficulty or agraphia may be part of visuo perceptual disorders when it is known as *spatial agraphia*. If the patient is unable to write because of inability to construct two dimensional figures, then it is known as *apraxic*. Difficulty in calculation or acalculia has two components, *spatial acalculia* is present when the patient is unable to place numbers properly and is associated with right parictal lesions, and difficulty in arithmetic is associated with left parietal lesions.
- **Pathology of Emotions:** The right parietal lobe is hypothesized to mediate emotions. Normally the right frontal lobe inhibits the right parietal lobe. Lesion of the right frontal lobe removes this inhibitory influence. Excessive emotions result as in mania or affective disorders. Electroencephalographic (ECG) recordings show the right hemisphere to be more active than the

left during emotional stimulation. Dysfunction of the right hemisphere is associated with affective disorders. Studies conducted on stroke patients have indicated that emotional processing is largely lateralized to the right hemisphere. Positive affect is mediated by the left hemisphere. Hence damage to the left hemisphere particularly the left frontal lobe is associated with depression. Negative affect is associated with the right hemisphere. Damage to the right hemisphere reduces the negative affective coning of experience. The result is a disproportionate positive affect resulting in mania. Normal emotions are the outcome of a balanced processing between the two hemispheres.

- **Sensory Template:** Polymodal association cortices of the cerebral cortex are present in the parietal lobes. It is connected to the frontal lobes and to the limbic system. Thus, it is hypothesized that the sensory template of the environment is constructed on a moment to moment basis by the parietal lobe. Its specialization for visuospatial attention spatial cognition as well as the polymodal association area enables it to scan the environment adequately and construct the sensory map. Thus while the frontal lobes are known as the executive centers of the brain the temporal lobes are known as the integrative centers of the brain the parietal lobes construct the sensory template of the world in a dynamic fashion.
- **Gerstmann's Syndrome:** Gerstmann's Syndrome is a collection of symptoms that occur as a result of a damaged parietal lobe prevailing in the left hemisphere. The patient has difficulty with writing, arithmetics, language, perception logics. It consists of acalculia, agraphia, finger agnosia and left-right disorientation.
- **Balint's Syndrome:** It diminishes motor skills simultaneously with visual attention. Voluntary eye movements may grow weak in the patient. These patients are not able to integrate all the attributes of a visual picture and may similarly be inadequate to manipulate an object in the inadequacy of a visual stimulus. If the right *parietal lobe damaged*, the strength to care for one's own body is diminished as the patient is incapable to notice for at least one side of the body. The ability to draw is similarly diminished.
- **Lack of awareness:** Like frontal lobe lesions, parietal lobe lesions are aalso often associated with a lack or reduction of awareness about one's impairments. In mild and moderate TBI patients with these lesions, lack of awareness is a common occurrence (Allen and Ruff, 1990; Sherer et al., 2003; Hart et al., 2006). Severity and time since injury are factors that influence this deficit. It has been reported that, patients with severe TBI show greater awareness deficits during the first 2 years after the injury. During this period, patients overestimate their performance and report more physical than cognitive

impairments. (Sherer et al., 1998; Vanderploeg et al., 2007). Importantly, awareness deficits can persist during the chronic stage of recovery (Prigatano, 2010; Wise et al., 2005). Individuals with right parietal and frontal lesions often suffer from neglect syndrome (Vallar and Perani, 1986; Verdon et al., 2010). This syndrome is defined as an attentional deficit of the contra-lesional space. In fact, patients with neglect become unaware of stimuli on their contra-lesional side (they may eat food from only the right side of the plate; ignore individuals on their left; miss words on the left side of the page). Even if neglect syndrome is often accompanied by lack of awareness, double dissociation between the two deficits have been reported (e.g., Bisiach et al., 1986). The lack of awareness is evaluated using self-report questionnaires, structured interviews and behavioural observations (Fleming et al., 1996; Sbordone et al., 2010; Fischer et al., 2004). Self-reported questionnaires, a commonly used method of assessment, compare responses given by patients with responses given by their clinician or caregiver. The discrepancy between patient responses and external evaluators provides evidence for the deficits. Pagulayan et al. (2007) investigated the magnitude of lack of awareness in a nonselect sample of patients with TBI. They found that levels of awareness varied both across cognitive domains and over time (at 1 month and 1 year post injury). To date, few studies have explored the longitudinal course of awareness, which is needed to further our understanding of this impairment. In order to better estimate the magnitude of reduced awareness across TBI patients, future studies should include patients who are not enrolled in rehabilitation programs.

OCCIPITAL LOBE: ANATOMY AND FUNCTIONS

The occipital cortex is located at the posterior part of the cerebral cortex. It lies above the cerebellum and is posterior to the parietal and temporal cortices. There are three Brodmann areas in this cortex. The most posterior of these is the area 17 which is also known as the striate cortex. The fibres of the optic tract coming from the eyes terminate here and give it the striate appearance to the naked eye. Area 17 is in the medial portion of the occipital cortex. It consists of simple cells which are sensitive to movement. These cells are also sensitive to specific location and orientation of lines. The cells fire when visual stimuli move. In addition, some of these cells fire when the stimuli are in a specific location and orientation. Colour perception also occurs in this area. Area 18 is adjacent to 17 laterally and contains complex cells. Each complex cell receives input from several simple cells. The complex cells are sensitive to orientation. However they are not location specific.

They fire across different locations if the orientation is maintained. Complex cell are predominantly present in area 18. Some of these cells receive converging input from both eyes, the remainder receive input only from one eye. The complex cells are probably involved in the earliest stages of actual form perception. Area 19 is adjacent to the area 18 laterally. It contains predominantly hyper complex cells which are sensitive to movement, position and orientation. They process angles, corners, and movements and analyze discontinuity. These cells process geometric forms and visual closure. This area also receives input from non visual areas of the brain such as the brain stem nuclei, the pontine arid mesencephalic reticular formation, the lateral amygdala and the lateral hypothalamus. These connections enable the processing of visual stimuli to be influenced by level of wakefulness and by emotional and motivational influences. The visual cortex is connected to the frontal regions. The association areas of the occipital cortex i.e., areas 18 and 19 arc connected to the inferior temporal lobe and to the parietal lobe. These connections take the processing of visual information beyond the occipital cortex. Visual "stimuli become positioned in space and personally meaningful through these connections.

OCCIPITAL LOBE SYNDROMES

Lesion in the occipital cortex is associated with various disorders of processing visual information. Large and bilateral lesions in the medial calcimine cortex or area 17 lead to cortical blindness. The patient is blind though the eye although the retina is intact. As this area receives maximum input from the fovea patterned vision is lost. The patient is able to discriminate levels of brightness and can make out light from dark. Peripheral vision medicated at the thalamic geniculate level would be present. Consequently though the patient is unable to see they do not bump into large objects. This phenomenon is known as blind sight. Probably because of this the patients with cortical blindness deny blindness. If confronted with their blindness, they give excuses for their disability. This is known as *Anton syndrome*. If the lesion is partial, the patient might have*hemi anopia* wherein loss of vision is present in only half of the visual field. Smaller lesions in the occipital lobe lead to hallucinations. Lesions in the area 17 result in hallucinations of moving lights, flashes, sparks and tongues of flame and colours. Objects become exceedingly large which is known as *macropsia* or exceedingly small which is known as *micropsia*. Objects also may be elongated or blurred in their outline. Colors might run and objects might lose colour. Lesions in the visual association areas i.e, areas 18 and 19 can produce complex visual hallucinations.

- **Agnosias:** are the other major disorders occurring in occipital lesions. *Visual objects agnosia*is the inability to recognise familiar objects through sight and is associated with medial occipital lesions, the patient is able to recognize the object though another sensory modality such as touch. Lesions of the left occipital lobe or disconnection of the occipital lobe from the left angular gyrus results in this agnosia. Sometimes, the patient is able to recognise the objects when the object is placed in a familiar context. The patient may be able to draw, paint or trace the object without recognizing it. The object becomes stripped of its meaning.
- **Optic Ataxia:** Optic ataxia is present when the object changes its appearance or disappears while the patient is looking at it. Inability to recognise colours or *colour agnosia* is another condition associated with inferior bilateral occipital lesions. The patient is unable to name, match and identify colour.
- **Prosopagnosia:** Or the inability to recognise familiar face by sight is associated with right occipital lesions. The patient recognizes a face as a face but is unable to place as to whom it belongs. He or she may also recognize the voice without recognizing the face. The lesions are in the right occipital and temporal regions.
- **Simultagnosia:** Is an inability to perceive the totality of a figure or a scene. The patient is unable to see the totality of the scene and can describe only parts of it. It is associated with difficulties in visual scanning. The ability to scan and visually explore the environment is drastically reduced. It is a beak down to perform visual serial step by step analysis. Visual attention is largely limited to the central visual field. The lesions are in the left hemisphere, or in the frontal eye fields. Bilateral superior occipital lesions can also cause this dysfunction.

CONCLUSION

The Parietal Lobe and the Occipital Lobe are the two of the four principal lobes of the brain. The parietal lobe is located behind the frontal lobes and above the temporal lobes. The functions of the parietal lobe include the appreciation of touch, awareness of the position of the extremities, vibratory sense and the fine tactile perception. Thus, the sensory cortex is contained within the parietal lobes. The sensory cortex is contained in the most forward part of the parietal lobes. This is just behind the *central sulcus*. This sensory cortex is on what is called the *post central gyrus* (ridge). Further to the rear in the parietal lobes are areas for the reception and organization of language functions. The parietal lobes integrate sensory data from other parts of the body. The parietal lobes have both the left and right hemispheres. The left side is concerned in symbolic functions in language and mathematics. The right hemisphere is more focused on images and spatial relationships.

The Occipital Lobes are located behind and beneath the Parietal Lobes and to the rear of the Temporal Lobes. The occipital lobes contain the primary visual cortex. The Occipital Lobes are divided into different functional visual areas. Each area contains a map of the visual world. The Occipital Lobes are the area where specialized neurons which receive and process visual information reside. The visual pathways run from the retina to the primary visual cortex, contained within the Occipital Lobes. While the eye is the lens for light and images that we see, it is the Occipital Lobes, in conjunction with other areas of the cerebral cortex, that actually see. The energy of the light is recorded on the retina, then transmitted through neurons with direct connections to the Occipital Lobes, where the meaning of the light pattern, is interpreted. Thus, it is possible to have a completely normal eye, but be blind, because of damage to the occipital lobes.

REFERENCES

Allen, C. C., & Ruff, R. M. (1990). Self-rating versus neuropsychological performance of moderate versus severe head-injured patients. *Brain Injury*, *4*(1), 7–17. doi:10.3109/02699059009026143 PMID:2297602

Bisiach, E., Vallar, G., Perani, D., Papagno, C., & Berti, A. (1986). Unawareness of disease following lesions of the right hemisphere: Anosognosia for hemiplegia and anosognosia for hemianopia. *Neuropsychologia*, *24*(4), 471–482. doi:10.1016/0028-3932(86)90092-8 PMID:3774133

Critchley, M. (1953). *The parietal lobes*. Williams and Wilkins.

Fischer, S., Trexler, L. E., & Gauggel, S. (2004b). Awareness of activity limitations and prediction of performance in patients with brain injuries and orthopedic disorders. *Journal of the International Neuropsychological Society*, *10*(2), 190–199. doi:10.1017/S1355617704102051 PMID:15012839

Fleming, J. M., Strong, J., & Ashton, R. (1996). Cluster analysis of self-awareness levels in adults with traumatic brain injury and relationshipto outcome. *The Journal of Head Trauma Rehabilitation*, *13*(5), 39–51. doi:10.1097/00001199-199810000-00006 PMID:9753534

Hart, S. J., Davenport, M. L., Hooper, S. R., & Belger, A. (2006). Visuospatial executive function in Turner syndrome: Functional MRI and neurocognitive findings. *Brain*, *129*(5), 1125–1136. doi:10.1093/brain/awl046 PMID:16504970

Pagulayan, K. F., Hoffman, J. M., Temkin, N. R., Machamer, J. E., & Dikmen, S. S. (2008). Functional Limitations and Depression After Traumatic Brain Injury: Examination of the Temporal Relationship. *Archives of Physical Medicine and Rehabilitation*, *89*(10), 1887–1892. doi:10.1016/j.apmr.2008.03.019 PMID:18929017

Prigatano, G. P. (2010). *Advances in the study of anosognosia*. New York: Oxford University Press.

Sbordone, R. J., Ronald, E., & Purisch, D. (2007). *Neuropsychology for Psychologists, Health Care Professionals and Attorneys*. CRC Press.

Sherer, M., Boake, C., Levin, E., Silver, B.V., Ringholz, G., & Walter, M.H. (1998). Characteristics of impaired awareness after traumatic brain injury. *Journal of the International Neuropsychology Society*, *4*, 380-387.

Sherer, M., Hart, T., & Todd, N. (2003). Measurement of impaired self-awareness after traumatic brain injury: A comparison of the patient competency rating scale and the awareness questionnaire. *Brain Injury: [BI]*, *17*(1), 25–37. doi:10.1080/0269905021000010113 PMID:12519645

Vallar, G., & Perani, D. (1986). The anatomy of unilateral neglect after right-hemisphere stroke lesions. A clinical/CT-scan correlation study in man. *Neuropsychologia*, *24*(5), 609–622. doi:10.1016/0028-3932(86)90001-1 PMID:3785649

Vanderploeg, R. D., Curtiss, G., Luis, C. A., & Salazar, A. M. (2007). Long-term morbidities following self-reported mild traumatic brain injury. *Journal of Clinical and Experimental Neuropsychology*, *29*(6), 585–598. doi:10.1080/13803390600826587 PMID:17691031

Verdon, V., Schwartz, S., Lovblad, K. O., Hauert, C. A., & Vuilleumier, P. (2010). Neuroanatomy of hemispatial neglect and its functional components: A study using voxel-based lesion-symptom mapping. *Brain*, *133*(3), 880–894. doi:10.1093/brain/awp305 PMID:20028714

Wise, K., Ownsworth, T., & Fleming, J. (2005). Convergent validity of self-awareness measures and their association with employment outcome in adults following acquired brain injury. *Brain Injury: [BI]*, *19*(10), 765–775. doi:10.1080/0269905050019977 PMID:16175837

KEY TERMS AND DEFINITIONS

Agnosia: Is the inability to process sensory information. Often there is a loss of ability to recognize objects, persons, sounds, shapes, or smells while the specific sense is not defective nor is there any significant memory loss.

Apraxias: Apraxia is a motor disorder caused by damage to the brain (specifically the posterior parietal cortex) in which the individual has difficulty with the motor planning to perform tasks or movements when asked, provided that the request or command is understood and the individual is willing to perform the task.

Colour Perception: Color perception is a part of the larger visual system and is mediated by a complex process between neurons that begins with differential stimulation of different types of photoreceptors by light entering the eye. Colour perception for humans is three dimensional.

Occipital Lobe: The occipital lobe is one of the four major lobes of the cerebral cortex in the brain of mammals. The occipital lobe is the visual processing center of the mammalian brain containing most of the anatomical region of the visual cortex.

Parietal Lobe: The parietal lobes are one of the four main lobes or regions of the cerebral cortex. The parietal lobes are positioned behind the frontal lobes and above the temporal lobes. These lobes are important to the function and processing of sensory information, understanding spatial orientation and body awareness.

Reading: Reading is the complex cognitive process of decoding symbols to derive meaning. It is a form of language processing. Success in this process is measured as reading comprehension. Reading is a means for language acquisition, communication, and sharing information and ideas.

Space Perception: Space perception is a process through which humans and other organisms become aware of the relative positions of their own bodies and objects around them. Space perception provides cues, such as depth and distance, that are important for movement and orientation to the environment.

Writing: Writing is the process of using symbols to communicate thoughts and ideas in a readable form. Writing is a medium of human communication that represents language with signs and symbols. For languages that utilize a writing system, inscriptions can complement spoken language by creating a durable version of speech that can be stored for future reference or transmitted across distance.

Chapter 10
Neuropsychological Profile of Various Neurological and Psychiatric Conditions

Ramprasad Santhanakrishnan K.
Dharwad Institute of Mental Health and Neurosciences, India

ABSTRACT

In the current chapter, the neuropsychological profile of various neurological and psychiatric conditions is focused on, including two major divisions (i.e., dementia—cortical and sub-cortical—and major mental disorders—substance use disorders, mood disorders, anxiety disorders, psychotic disorders, sleep disorders, childhood disorders, personality disorders, and sexual disorders). Both divisions have sub-classifications that include introduction, etiopathogenesis, epidemiology, clinical features, evaluation, treatment, and psychosocial aspects.

INTRODUCTION

Dementia is a progressive impairment in the functions of the brain especially of the cognitive domain when there is clear conscious. This leads to decline in the previously learned things in more than one domain's executive function, attention, language, memory, complex motor tasks, following social norms. These will lead to decline in the level of functioning. Dementia is derived from a Latin word Dementatus meaning out of one's mind (Sadock, Kaplan & Sadock, 2005). It was first used by Celsius in 1st century AD, till 19th century there was no much knowledge (Jellinger, 2010). Esquirol also described hallucinations, delusions, and aggressive behavior motor

DOI: 10.4018/978-1-7998-2860-0.ch010

Figure 1. Insoluble fibrillar Aβ plaques begins in neocortex

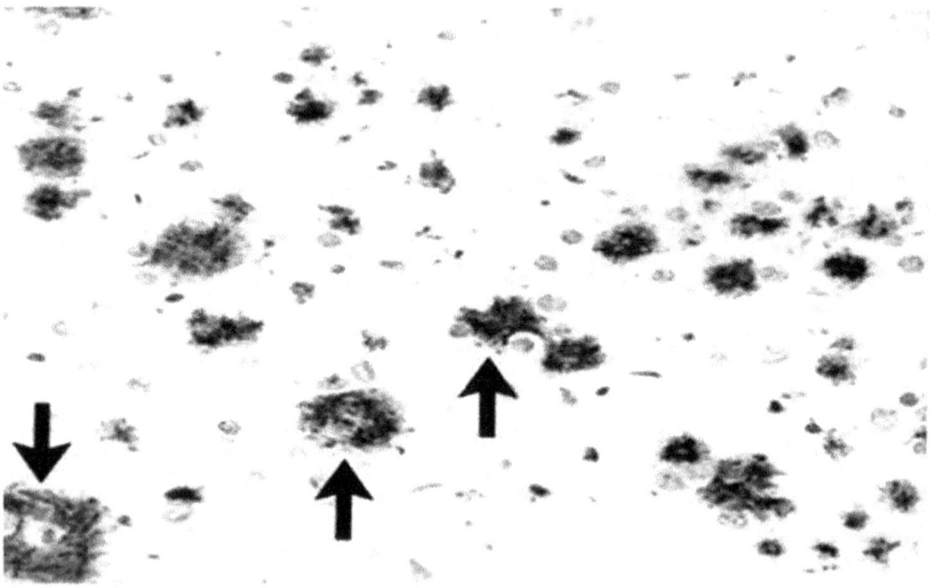

impairments in dementia patients. In 1907 Alois Alzheimer was the first person to identify histopathological changes associated with progressive degenerative dementia.

- **Etiopathogenesis:** The etiopathogenesis are diverse including degenerative disorders, medical illness, infections, substance induced and combination of these. Histopathologically the presence of neurofibrillary tangles and plagues in the cortex and brainstem; Fig: 1, this is seen in 42% of patients with Alzheimer dementia, Lewy Body dementia is seen in 15 to 35 percent (Kotzbauer, Trojanowsk, & Lee, 2001). The second most common dementia and most common below the age of 65 is fronto-temporal degenerative dementia. The causes of dementia are Neurodegenerative, Vascular, Neurological diseases, Endocrine, Nutritional, infectious, Metabolic, Traumatic, Exposure to substances like Alcohol, Irradiation. Minimum cognitive impairment - amounts to discrete loss of cognitive functions but not meeting the criteria of the dementia disorder as such. This also has some mood influences also.

ICD - 10 Criteria for Diagnosis of Dementia

1. Presence of a dementia symptoms and signs.
2. Insidious onset with slow deterioration. While the onset usually seems difficult to pinpoint in time, realization by others that the defects exist may come suddenly. An apparent plateau may occur in the progression.
3. Absence of clinical evidence, or findings from special investigations, to suggest that the mental state may be due to other systemic or brain disease which can induce a dementia (e.g. hypothyroidism, hypercalcaemia, vitamin B12 deficiency, niacin deficiency, neurosyphilis, normal pressure hydrocephalus, or subdural haematoma).
4. Absence of a sudden, apoplectic onset, or of neurological signs of focal damage such
5. As hemiparesis, sensory loss, visual field defects, and incoordination occurring early in the illness (although these phenomena may be superimposed later).

ICD-10 Criteria for Delirium

1. Impairment of consciousness and attention (on a continuum from clouding to coma; reduced ability to direct, focus, sustain, and shift attention);
2. Global disturbance of cognition (perceptual distortions, illusions and hallucinations most often visual; impairment of abstract thinking and comprehension, without transient delusions, but typically with some degree of incoherence; impairment of immediate recall and of recent memory but with relatively intact remote memory; disorientation for time as well as, in more severe cases, for place and person);
3. Psychomotor disturbances (hypo- or hyperactivity and unpredictable shifts from one to the other; increased reaction time; increased or decreased flow of speech; enhanced startle reaction);
4. Disturbance of the sleep-wake cycle (insomnia or, in severe cases, total sleep loss or reversal of the sleep-wake cycle; daytime drowsiness; nocturnal worsening of symptoms; disturbing dreams or nightmares, which may continue as hallucinations after awakening);
5. Emotional disturbances, e.g. depression, anxiety or fear, irritability, euphoria, apathy, or wondering perplexity. The onset is usually rapid, the course diurnally fluctuating, and the total duration of the condition less than 6 months.

In a certain proportion of cases, the features of Alzheimer's disease and vascular dementia may both be present. In such cases, double diagnosis (and coding) should be made. When the vascular dementia precedes the Alzheimer's disease, it may be impossible to diagnose the latter on clinical grounds.

- **Differential Diagnosis:** A depressive disorder (F30-F39); delirium (F05.-); organic amnesic syndrome (F04); other primary dementias, such as in Pick's, Creutzfeldt-Jakob or Huntington's disease (F02.-); secondary dementias associated with a variety of physical diseases, toxic states, etc. (F02.8); mild, moderate or severe mental retardation (F70-F72). (The ICD-10, 1993).

Neurocognitive Impairment

1. Cortical dementia
2. Subcritical dementia

Mixed signs and symptoms are commonly seen i.e., both cortical and subcortical signs and symptoms are see together, hence designation like this is not routinely used,; instead use of Cortical dementia, Parkinson dementia, vascular dementia, Dementia of Lewy body are designated.

- **Epidemiology:** Incidence of 1.9% in the age group of 65-70 yrs; 6.3% in the age group of 70-80 yrs; 21.7% in the age group of 85-90 yrs; 47.5% in the age above 95 yrs. is been seen. 35.6 million people with dementia are been identified till 2010 and its increase to 115.4 million by 2050 (Liz, 2016). The prevalence varies from region to region.
- **Risk and Protective Factors:** Leading risk factor is old age. Women has increased chances of getting Dementia as their life expectancy is high than men. Vascular risk factors for CVA have increased risk for dementia for Vascular, Alzheimer or dementia in general. Vascular risk factors in midlife are: increased BMI, High Cholesterol. Hypotension in old age and long standing depression burden in young life are associated with increased risk of developing Dementia and/or Minimal Cognitive Impairment. Cerebrovascular accident, Cardio vascular accident, Congestive cardiac failure, Atrial fibrillation Obstructive sleep apnea syndrome are all having increased risk for dementia. Nicotine use tends to increase risk of Alzheimer disease, Vascular dementia, Parkinson's disease also. Formal education tends to postpone the onset of dementia into later life.

Estrogen replacement therapy in menopause women has been seen to protect them from developing dementia.

Head injury and or infection to the brain parenchyma also results in dementia for various degree and the deficits depend upon the structures involved.

- **Clinical Features:** It varies depending on the site of involvement. If there is Cognitive involvement it will affect the Personal hygiene care totally, moving about, thinking processes, working patterns, feeding fads, confusions, forgetfulness, loss of learned skills / previously known works are not done appropriately. Functionally impaired depends on the extent of the motor involvement or the sensory involvement or the association areas involved [acknowledgement, processing and execution all are involved in this]. This is seen as difficulties in job performances, managing finances, managing medicines, activities of daily living, trouble in maintaining social relationship and activities. They also have increased risk of developing Adverse drug reactions, and mood disorders especially Depression.
- **Screening:** it depends on the medical setting the person has attended. Hus need for the family and his personal need of the functions. Examples: Need of motor skills are needed for labourers. Cognitive skills are needed for office works.
- **Evaluation:** History by the patient has to be taken and is at most important. The attenders who are along with him living with him/her having constant interaction with them are interviewed to know the extent of the problem of the illness for the people around the patient-burden on the caregivers (Reinhard, Given, Petlick, & Bemis, 2008). It may be in the *Onset; Evolution; Deterioration.*

All insults may be mild to severe in severity which is congruent with the severity of dementia. Bedside cognitive tests: Montreal Cognitive Assessment. It is a 30 point test can be administered in 10 minutes. It has 90% sensitivity for Minimal cognitive Impairment and 87% specificity for evaluation of cognition

Severity is measured by Cognitive Dementia Rating Scales [CDR]- it measures the Orientation, Memory, Time and Location, Personal care judgment, problem solving, community affect. It is the Functional Assessment and Staging [FAST].

- **Neuropsychiatric Manifestations:** The deficiencies in the Dementia leads to mood changes most commonly Depression. Anxiety are also seen in some, in long standing persons we can see the Personality change may be due to symptoms or the effect of the disorder. Emergence of Psychotic symptoms can occur, Sleep disturbances, Agitation and/or aggression, again due to the mood changes or because to the disorder are commonly seen. Alzheimer disease affects posterior Cingulate gyrus, Temporal and association cortices (Julie, Graziane., Robert., & Sweet, 2017).

- **Laboratory Investigations:** Commonly used investigations are routine blood examinations, Harmone levels, neuroimaging like CT scan, MRI, PET, FDGPeT scan CSF analysis- Aβ protein, Tau protein, Phospho Tau increased in Alzheimer.
- **Neuropsychiatry Tests:** Depending upon the need different lobe tests are done. This helps in the treatment and prognosis of the patient.

Treatment

1. Pharmacology
2. Non Pharmacological (Psychosocial)
    - **Pharmacological Treatment:** The most common neurotransmitter implicated in the processes of dementia is deficiency of acetylcholine in the brain at the synaptic level. In this background the supplementation of the same is done by giving centrally acting acetylcholineesterase inhibitors to decrease the degradation of the released acetylcholine neurotransmitter at the synaptic cleft. Eg: Donepezil, Rivastigmine & Galantamine. Memantine - a noncompetitive NMDA receptor antagonist-modulates excitation and have neuroprotective action is also used.
    - **Psychosocial aspects:** It confers to treatment other than Pharmacological way of dealing the symptoms of the Disease.

Nonpharmacological Treatment

1. **Psychotherapy:** to the patient has limitations but for the caregivers is important
2. **Psychosocial:** addressing the social issues related to the disorder.
3. **Behavioral**: the effect of the behavior of the patient towards the members of family and neighbours, for any other transactions.
4. Cognitive retraining
5. Stimulation and reorientation therapy

Specific Dementias

1. Alzheimer dementia
2. Frontotemporal dementia
3. Corticobasal degeneration
4. Progressive supraneuclear palsy
5. Dementia with Lewy bodies
6. Dementia in Parkinson disease

7. Dementia in Huntington disease
8. Prion disease
   - **Pathology in Different Types of Dementia:** In Alzheimer dementia the main pathology is macroscopically there is atrophy of the hippocampus. Microscopically there are amayloid plaques and neurofibrillary tangles. Amyloid plaques are deposits of insoluble fibrillar Aβ protein in Neocortex, Entorhinal cortex and hippocampus, subcortical regions. The neurofibrillary tangles are intracellular inclusion bodies containing paired helical filaments composed of aggregates of hyperphosphorylated Microtubule-associate protein-t [MAPT]

Aggregation of α-synuclein, a fibrillary protein as intraneuronal inclusions, [Lewy bodies] seen in 50% of Alzheimer dementia in amygdala, transentorhinal cortex.

Transactive response DNA-binding protein of 43 kDa [TDP-43], most frequently aggregated protein in Frontotemporal lobar degeneration. Usually this will be present in the mesial temporal lobe and to a lesser degree in the Neocortex.

Other pathology associated is – loss of neurons, in different areas of brain mainly Entorhinal cortex and hippocampus. Later in the disease loss of cholinergic neurons in nucleus basalis and loss of noradrenergic and serotonergic neurons in the brainstem is seen. Granulovascular degeneration with Hirano bodies affects the Pyramidal neurons at Hippocampus (Vyas, & Niraj Ahuja, 2011).

Prion diseases: They include Kuru, CrutzfieldJacobs disease, Gerstmann–Straussler– Scheinker (GSS) disease and fatal insomnia.

MAJOR MENTAL DISORDERS

1. Substance use disorders
2. Mood disorders
3. Anxiety disorders
4. Psychotic disorders
5. Sleep disorders
6. Childhood disorders
7. Personality disorders
8. Sexual disorders

Further major mental disorders are elucidated in following order

SUBSTANCE USE DISORDERS

- Introduction
- Etiopathogenesis
- Epidemiology
- Clinical features
- Evaluation
- Treatment
- Psychosocial aspects

Substances are those things which are used by people in the community and are having effect over the body / brain for decreasing the distress in their living. It amounts to reducing the distress to the body / mind. Most commonly used substances are Nicotine, Alcohol, Opioids, Pain killers, Cannabis, CNS Stimulants, Coffee, Tea and others. These substances are legally permitted to be used in some countries; hence the use may be legally offensive in some countries where it is not legally permitted. There is no sex discrimination in the usage of substances. At times it may symbolize the economic status of the person. In recent times influence of western culture on the eastern culture has started new entity social drinking.

- **Substance:** It states that the thing which are having their effect on the central Nervous System neurons giving pleasant feeling and is helping the person forget the undesired situations/ things which they cannot forget, experiencing different feelings which are attained when the substance are consume
- **Use:** It is a term which states that the substance is used.
- **Abuse:** It means that the substance use perse or its effect is not acceptable to the person or to the surrounding people.
- **Tolerance:** It is a state or condition where the usual amount of the substance use is not giving the desired effect which the person wants or the effect which the substance had given during its initial use. In order to get the desired effect they consume more amount of the substance.
- **Intoxication:** Acute intoxication is a transient phenomenon. Intensity of intoxication lessens with time, and effects eventually disappear in the absence of further use of the substance. Recovery is therefore complete except where tissue damage or another complication has arisen.
- **Dependence:** It is of two types
    1. Psychological dependence
    2. Physical dependence

- **Psychological Dependence**: It states that the person has longing for the substance used and is agitated or perturbed/ disturbed without the substance. He/she would do what so ever to procure the substance and consume it.
- **Physical Dependence:** It states that the person would have one or the other physical symptoms if they do not consume the substance in time or appropriate amount needed for them. They may be agitation, anxiousness, tremors, sweating, increased heartbeat, increased respiratory rate, frequent urination, pain abdomen, jitteriness, seizure.

The person would do something to procure the substance in order to avoid these symptoms which are not acceptable to anyone. They have their first priority to procure and consume the substance and then the other works.

A definite diagnosis of dependence should usually be made only if three or more of the following have been present together at some time during the previous year:

1. A strong desire or sense of compulsion to take the substance;
2. Difficulty in controlling substance-taking behaviour in its onset, termination, or
3. levels of use;
4. Aphysiological state when substance use has ceased or reduced, as evidenced by: the
5. characteristic withdrawal syndrome for the substance; or use of the same or related
6. Substance to relieve or avoid withdrawal symptoms;
7. Evidence of tolerance, such that increased doses of the psychoactive substance are required in order to achieve effects originally produced by lower doses (clear examples of this are found in alcohol- and opiate-dependent individuals who may take daily doses sufficient to incapacitate or kill non tolerant users);
8. Progressive neglect of alternative pleasures or interests because of
9. Psychoactive substance use, increased amount of time necessary to obtain or take the substance or to recover from its effects;
10. Persisting with substance use despite clear evidence of overtly harmful consequences, Such as harm to the liver through excessive drinking, depressive mood states consequent to periods of heavy substance use, or drug-related impairment of cognitive functioning; efforts should be made to determine that the user was actually, or could be expected to be, aware of the nature and extent of the harm.
    - **Etiopathogenesis:** Alcohol use is influenced by mainly two factors, Genetic control and Social control. These may have influence in starting, maintaining and relapse of the use disorder. Other substances use is

related to availability and peers pressure. Personality of a person has a major influence on these activities.

- **Clinical features:** ICD-10
- **Withdrawal state:** A group of symptoms of variable clustering and severity occurring on absolute or relative withdrawal of a substance after repeated, and usually prolonged and/or high-dose, use of that substance. Onset and course of the withdrawal state are time-limited and are related to the type of substance and the dose being used immediately before abstinence. The withdrawal state may be complicated by convulsions.
- **Diagnostic Guidelines:** Withdrawal state is one of the indicators of dependence syndrome and this latter diagnosis should also be considered. Withdrawal state should be coded as the main diagnosis if it is the reason for referral and sufficiently severe to require medical attention in its own right. Physical symptoms vary according to the substance being used. Psychological disturbances (e.g. anxiety, depression, and sleep disorders) are also common features of withdrawal. Typically, the patient is likely to report that withdrawal symptoms are relieved by further substance use. It should be remembered that withdrawal symptoms can be induced by conditioned/learned stimuli in the absence of immediately preceding substance use. In such cases a diagnosis of withdrawal state should be made only if it is warranted in terms of severity.
- **Differential Diagnosis:** Many symptoms present in drug withdrawal state may also because by other psychiatric conditions, e.g. anxiety states and depressive disorders. Simple "hangover" or tremor due to other conditions should not be confused with the symptoms of a withdrawal state.

The diagnosis of withdrawal state may be further specified by using the following five-character codes: *Uncomplicated with convulsions*

Complicated by delirium

Alcohol-induced *delirium tremens* should be coded here. *Delirium tremens* is a short-lived, but occasionally life-threatening, toxic-confusional state with accompanying somatic disturbances. It is usually a consequence of absolute or relative withdrawal of alcohol in severely dependent users with a long history of use. Onset usually occurs after withdrawal of alcohol. In some cases the disorder appears during an episode of heavy drinking, in which case it should be coded here.

Prodromal symptoms typically include insomnia, tremulousness, and fear. Onset may also be preceded by withdrawal convulsions. The classical triad of symptoms includes clouding of consciousness and confusion, vivid hallucinations and illusions affecting any sensory modality, and marked tremor. Delusions, agitation, insomnia or sleep-cycle reversal, and autonomic over activity are usually also present.

- **Excludes:** Delirium, not induced by drugs and alcohol
- **Treatment:** First Alcohol is substituted by benzodiazepines for the detoxifying processes along with the replacement of the B_1 – Thiamine vitamin under the supervision of the Psychiatrist and General Physician.

The second part is the maintenance of the abstinence by modifying the trigger factors and maintenance factors for the substance use through behavioural, cognitive behavioural therapy.

- **Psychosocial Aspects:** The social stigma attached to the consumption of the substance has lead to the identification of the individual or their family in terms of the substance used and has become known to others in the society. They are kept away from the other social activities of that locality. To reduce these effects abstinence and remodeling of their behavior to be accepted in their society is very much needed (Vyas et al., 2011).

MOOD DISORDERS

Mood disorders are very common, which ranges from sadness to happy mood. The severity varies from time to time, in the same person and in different persons in the same situations. Mood disorders are found to run in family, and the risk is increased among the first degree relatives 8-9% compared to general population. The major neurotransmitter involved in the pathogenesis is found to be serotonin, norepinephrine, and acetylcholine. Metabolism of this catecholamine is basically done by mono-amino-oxidase inhibitors A and B. Recently glutamine activity on the NMDA receptors in the brain is also implicated. Prevalence of depression is increasing by many folds in the recent past, because of the environmental factors such as stress in the life at various places like schooling, working and household activities. This results in conflicts of inter-personal and intra-personal relations. WHO has found that the global burden by depression is increasing and by 2025 it is found to be the most prevalent disorder in the population. Females have an increased tendency to develop depression about 5% more than that of males. It can occur at any time but more commonly in the second decade. Hormonal parameters also have its influence in the development of mood disorders; Estrogen supplementation in the post-menopausal period has found to be protective in developing depression in females (Whooley, Grady., & Cauley, 2000).

- **Clinical Features:** The most common symptom expressed by the person is sadness, hypo activity, decreased responsiveness to happiness of self and others, decreased interest in day to day activities or perceiving new things. These occur due to the cognitive determinants of hopelessness, helplessness and worthlessness. The severity varies from mild, moderate to severe forms. With the above said cognition, the person may go into an attempt of ending his life (suicide).

Most severe form of depression is called as melancholia. It was described by Hippocrates. In this the vegetative symptoms are very severe which may lead to confusion with psychosis. The first episode duration of 2 weeks with the above symptoms mandate the diagnosis.

Dysthymia is mild variety of depression which is pulled through in the life of the person, along with normal day to day activities, for a period of more than 2 years.

The dexamethasone suppression test is impaired in depression.

Mania is the other end of mood spectrum. Happiness is the center of mood symptoms. Symptoms are happiness, decreased sleep followed by no tiredness after getting up from the sleep, hyperactivity which is goal-directed; the thoughts of believing that he has high or supernatural powers as compared to others, Grandiosity is commonly seen. Other symptoms like singing songs, helpful nature to others, other constructional activities which are not completed are commonly seen. The first episode duration of 1 week with the above symptoms mandate the diagnosis.

Hypomania is a condition where the person meets all the requirements of mania with a duration being less than 1 week. The person most of the times will be smiling and responds warmly with the surrounding people, hence many a times this is passed unrecognized.

Psychosis is a condition where the person is not aware of the day and night discrimination, lack of personal hygiene, in full consciousness. Hallucinations in either modality, most commonly auditory and visual are seen. This can occur along with mood disorders. This occurs when there is increase in the neurotransmitter dopamine at D2 receptors in the brain (American Psychiatric Association, 2013).

Spectrum of Mood Disorders

Paranoid Spectrum					
Cluster A Personality Disorders	Paranoia	Paraphrenia	Paranoid Schizophrenia	Other Schizophrenia	Dementia with Paranoid features

ICD 10

Three degrees of severity are specified here, sharing the common underlying characteristics of elevated mood, and an increase in the quantity and speed of physical and mental activity. All the subdivisions of this category should be used only for a single manicepisode. If previous or subsequent affective episodes (depressive, manic, or hypomanic), the disorder should be coded under bipolar affective disorder (ICD-10,1993).

Hypomania

Hypomania is a lesser degree of mania, in which abnormalities of mood and behaviour are too persistent and marked to be included under cyclothymia, but are not accompanied by hallucinations or delusions. There is a persistent mild elevation of mood (for at least several days on end), increased energy and activity, and usually marked feelings of well-being and both physical and mental efficiency. Increased sociability, talkativeness, over familiarity, increased sexual energy, and a decreased need for sleep are often present but not to the extent that they lead to severe disruption of work or result in social rejection. Irritability, conceit, and boorish behaviour may take the place of the more usual euphoric sociability. Concentration and attention may be impaired, thus diminishing the ability to settle down to work or to relaxation and leisure, but this may not prevent the appearance of interests in quite new ventures and activities, or mild over-spending.

- **Diagnostic Guidelines:** Several of the features mentioned above, consistent with elevated or changed mood and increased activity should be present for at least several days on end, to a degree and with persistence greater than described for cyclothymia. Considerable interference with work or social activity is consistent with a diagnosis of hypomania, but if disruption of these is severe or complete, mania should be diagnosed.

Mood is elevated out of keeping with the individual's circumstances and may vary from carefree joviality to almost uncontrollable excitement. Elation is accompanied by increased energy, resulting in overactivity, pressure of speech, and a decreased need for sleep. Normal social inhibitions are lost, attention cannot be sustained, and there is often marked distractability. Self-esteem is inflated, and grandiose or over-optimistic ideas are freely expressed

- **Bipolar Affective Disorder:** This disorder is characterized by repeated (i.e. at least two) episodes in which the patient's mood and activity levels are significantly disturbed, this disturbance consisting on some occasions of an elevation of mood and increased energy and activity (mania or hypomania), and on others of a lowering of mood and decreased energy and activity(depression). Characteristically, recovery is usually complete between episodes, and the incidence in the two sexes is more nearly equal than in other mood disorders. As patients who suffer only from repeated episodes of mania are comparatively rare, and resemble (in their family history, premorbid personality, age of onset, and long-term prognosis) those who also have at least occasional episodes of depression, such patients are classified as bipolar. Manic episodes usually begin abruptly and last for between 2 weeks and 4-5 months (median duration about 4 months). Depressions tend to last longer (median length about 6 months), though rarely for more than a year, except in the elderly. Episodes of both kinds often follow stressful life events and other mental trauma, but the presence of such stress is not essential for the diagnosis. The first episode may occur at any age from childhood to old age. The frequency of episodes and the pattern of remissions and relapses are both very variable, though remissions tend to get shorter as time goes on and depressions to become commoner and longer lasting after middle age. Although the original concept of "manic-depressive psychosis" also included patients who suffered only from depression, the term "manic-depressive disorder or psychosis" is now used mainly as a synonym for bipolar disorder. It includes: manic-depressive illness, psychosis or reaction.

- **Depressive Episode:** In typical depressive episodes of all three varieties mild, moderate and severe, the individual usually suffers from depressed mood, loss of interest and enjoyment, and reduced energy leading to increased fatigability and diminished activity. Marked tiredness after only slight effort is common. Other common symptoms are:
    1. Reduced concentration and attention;
    2. Reduced self-esteem and self-confidence;
    3. Ideas of guilt and unworthiness (even in a mild type of episode);
    4. Bleak and pessimistic views of the future;
    5. Ideas or acts of self-harm or suicide;
    6. Disturbed sleep
    7. Diminished appetite.

The lowered mood varies little from day to day, and is often unresponsive to circumstances, yet may show a characteristic diurnal variation as the day goes on. As with manic episodes, the clinical presentation shows marked individual variations, and a typical presentations are particularly common in adolescence. In some cases, anxiety, distress, and motor agitation may be more prominent at times than the depression, and the mood change may also be masked by added features such as irritability, excessive consumption of alcohol, histrionic behaviour, and exacerbation of pre-existing phobic or obsessional symptoms, or by hypochondriacal preoccupations. For depressive episodes of all three grades of severity, a duration ofat least 2 weeks is usually required for diagnosis, but shorter periods may be reasonable if symptoms are unusually severe and of rapid onset.

Recurrent Depressive Disorder

- **Evaluation:** Basic biochemical parameters have to be examined to rule out other co-existing medical disorders. Specific investigations include thyroid function tests, Dexamethasone suppression test.

Psychological Tests

- **Depression:** Hamilton depression rating scale, Beck's depression inventory, Mental status examination
- **Mania:** Young Mania Rating Scale

Mental Status Examination

- **Treatment for Depression:** Mood elevators like Tri-cyclic antidepressants, eg: Amitryptylline, Imipramine, MAO inhibitors, eg: Selegilline, Selective Serotonin Reuptake Inhibitors, eg: Escitalopram, Norepinephrine reuptake inhibitors, eg: Trazadone, Acetylcholine reuptake inhibitors, eg: Mirtazepine, Combinations of the above, NMDA receptor antagonists, eg: Gabapentin
- **Treatment for Mania:** Mainly antipsychotics are used. Olanzepine, Haloperidol, Clonazepam are said to have anti-manic properties.

A person who has one disorder of the spectrum either depression or mania, tend to have the other spectrum in their lifetime, they are designated as bipolar illness. First episode mania getting depression in their life time is about 30-40%. First episode depression getting mania in their lifetime is about 15-20%. Patient also can have recurrent same episodes either depression or mania, they are designated as unipolar disorders. It has been found that mood stabilization prevents the person from switching from one form to the other.

- **Mood Stabilizer:** Gold standard drug with narrow therapeutic index is Lithium.

Others are

1. Sodium valproate
2. Olanzepine 10mg + Fluoxetine 20mg
3. Carbamazepine
4. Lamotrigine
    - **Psychosocial aspects:** Depression being the most common disorder has major impact in the output of the person. It has been found that there is 50-80% decrease in the working output in depressed patients. This in turn effects on the economy of the family and in turn burden to the country. Early recognition and in time treatment prevents this. This can be achieved through the psychiatric social work people. Family education regarding the illness in the family member, its early recognition, to enhance the compliance can be done. Community awareness helps in early detection and treatment.

ANXIETY DISORDERS

It includes:

Panic attacks
Panic disorder
Specific phobias
Generalized anxiety disorder
Obsessive compulsive disorder

Anxiety is a state of mind where there is activation of the Autonomic Nervous System. The symptoms and signs are that of the hyperarousal due to increase in the hormones like Adrenaline, Noradrenaline. The symptoms are increased awareness of one's own heartbeat which is racing and fearful, increased respiratory rate, sweating, perspiration, dry mouth, piloerection, frequent urination and or defecation, dilated pupils. The deficiency of serotonin neurotransmitters in different parts of the brain has been implicated.

- **Etiopathogenesis:** The precipitation of anxiety is multimodal like biological, social and the immediate influence of the day to day happenings.
- **Epidemiology:** The prevalence of anxiety in various presentations is increasing day by day. The cause could be the changing dynamicity of the life, perceiving the life activities as stressors, individualization and autonomy, inappropriate decisions taken at inappropriate lifetimes.

- **Clinical Features:** The pathognomonic and central signs and symptoms could be hyper-activation of the autonomic nervous system of varying severity in the individuals. The stimulus for the arousal of the autonomic nervous system varies across the spectrum of anxiety disorders.
- **Agoraphobia:** Fear of local spaces; Claustrophobia – Fear of closed spaces; Acrophobia – Fear of heights; Hydrophobia- Fear of water (excluding Rabies); Pyrophobia - Fear of fire; Arachnophobia – Fear of spiders; Xenophobia- Fear of people from different countries, regions; Phobophobia – Fear of fear
- **Panic Attacks:** The attack comprises of hyper-arousal of the autonomic nervous system with fear, escaping from the fear stimuli; with motor behaviors like running away most of the times. The period usually starts and reaches its peak within 5-10 minutes, lasts for half an hour to one hour and thereafter the intensity decreases gradually over a period of 45 minutes to an hour; the whole episode lasting for 1.5 to 2 hours. This is the usual episode of panic attacks; however typical descriptive attacks are less commonly seen. Many persons experience the panic/anxiety attacks for a brief period of time, say about few minutes and recover completely.

Panic attacks occurring regularly for more than 2 years is named as panic disorder.

The anxiety symptoms overwhelming the person, disrupting the usual daily life activities and is present throughout the day except during the sleep time is named as generalized anxiety disorder

ICD 10

In this group of disorders, anxiety is evoked only, or predominantly, by certain well-defined situations or objects (external to the individual) which are not currently dangerous. As a result, these situations or objects are characteristically avoided or endured with dread. Phobic anxiety is indistinguishable subjectively, physiologically, and behaviorally from other types of anxiety and may vary in severity from mild unease to terror. The individual's concern may focus on individual symptoms such as palpitations or feeling faint and is often associated with secondary fears of dying, losing control, or going mad. The anxiety is not relieved by the knowledge that other people do not regard the situation in question as dangerous or threatening. Mere contemplation of entry to the phobic situation usually generates anticipatory anxiety. The adoption of the criterion that the phobic object or situation is external to the subject implies that many of the fears relating to the presence of disease (nosophobia) and disfigurement (dysmorphobia) are now classified under hypochondriacal disorder. However, if the fear of disease arises predominantly and repeatedly from possible exposure to infection or contamination, or is simply a fear of medical procedures (injections, operations, etc.) or medical establishments (dentists' surgeries, hospitals, etc.) (ICD-10, 1993).

Phobic anxiety often coexists with depression. Pre-existing phobic anxiety almost invariably gets worse during an inter current depressive episode. Some depressive episodes are accompanied by temporary phobic anxiety and a depressive mood often accompanies some phobias, particularly agoraphobia. Whether two diagnoses, phobic anxiety and depressive episode, are needed or only one is determined by whether one disorder developed clearly before the other and by whether one is clearly predominant at the time of diagnosis. If the criteria for depressive disorder were met before the phobic symptoms first appeared, the former should be given diagnostic precedence. Most phobic disorders other than social phobias are more common in women than in men.

Panic Disorder (Episodic Paroxysmal Anxiety)

The essential features are recurrent attacks of severe anxiety (panic) which are not restricted to any particular situation or set of circumstances, and which are therefore unpredictable. As in other anxiety disorders, the dominant symptoms vary from person to person, but sudden onset of palpitations, chest pain, choking sensations, dizziness, and feelings of unreality (depersonalization or derealization) are common. There is also, almost invariably, a secondary fear of dying, losing control, or going mad. Individual attacks usually last for minutes only, though sometimes longer; their frequency and the course of the disorder are both rather variable. An individual in a panic attack often experiences a crescendo of fear and autonomic symptoms which results in an exit, usually hurried, from wherever he or she maybe. If this occurs in a specific situation, such as on a bus or in a crowd, the patient may subsequently avoid that situation. Similarly, frequent and unpredictable panic attacks produce fear of being alone or going into public places. A panic attack is often followed by a persistent fear of having another attack.

- **Diagnostic Guidelines**: In this classification, a panic attack that occurs in an established phobic situation is regarded as an expression of the severity of the phobia, which should be given diagnostic precedence. Panic disorder should be the main diagnosis only in the absence of any of the phobias. For a definite diagnosis, several severe attacks of autonomic anxiety should have occurred within a period of about 1 month:
    1. In circumstances where there is no objective danger;
    2. Without being confined to known or predictable situations; and
    3. With comparative freedom from anxiety symptoms between attacks (although anticipatory anxiety is common).

- **Generalized Anxiety Disorder:** The essential feature is anxiety, which is generalized and persistent but not restricted to, or even strongly predominating in, any particular environmental circumstances (i.e. it is "free-floating"). As in other anxiety disorders the dominant symptoms are highly variable, but complaints of continuous feelings of nervousness, trembling, muscular tension, sweating, lightheadedness, palpitations, dizziness, and epigastric discomfort are common. Fears that the sufferer or a relative will shortly become ill or have an accident are often expressed, together with a variety of other worries and forebodings. This disorder is more common in women, and often related to chronic environmental stress. Its course is variable but tends to be fluctuating and chronic.
- **Diagnostic Guidelines:** The sufferer must have primary symptoms of anxiety most days for at least several weeks at a time, and usually for several months. These symptoms should usually involve elements of:
    1. Apprehension (worries about future misfortunes, feeling "on edge", difficulty in concentrating, etc.);
        a. Motor tension (restless fidgeting, tension headaches, trembling, inability to relax); and
        b. Autonomic overactivity (lightheadedness, sweating, tachycardia or tachypnoea,
        c. Epigastric discomforts, dizziness, dry mouth, etc.).

In children, frequent need for reassurance and recurrent somatic complaints may be prominent. The transient appearance (for a few days at a time) of other symptoms, particularly depression, doesnot rule out generalized anxiety disorder as a main diagnosis, but the sufferer must not meet the full criteria for depressive episode, phobic anxiety disorder, panic disorder, or obsessive-compulsive disorder.

- **Obsessive-Compulsive Disorder:** The essential feature of this disorder is recurrent obsessional thoughts or compulsive acts. (Forbrevity, "obsessional" will be used subsequently in place of "obsessive-compulsive" when referring to symptoms.) Obsessional thoughts are ideas, images or impulses that enter the individual's mind again and again in a stereotyped form. They are almost invariably distressing (because they are violent or obscene or simply because they are perceived assenseless) and the sufferer often tries, unsuccessfully, to resist them. They are, however, recognized as the individual's own thoughts, even though they are involuntary and often repugnant. Compulsive acts or rituals are stereotyped behaviours that are repeated again and again. They are not inherently enjoyable, nor do they result in the completion of inherently useful tasks. The individual often views them as preventing some objectively unlikely event, often involving harm to or caused by himself or herself. Usually, though not invariably, this behaviour is recognized by the individual as pointless or ineffectual and repeated attempts are made to resist it; in very long-standing cases, resistance may be minimal. Autonomic anxiety symptoms are often present, but distressing feelings of internal or psychic tension without obvious autonomic arousal are also common. There is a close relationship between obsessional symptoms, particularly obsessional thoughts, and depression. Individuals with obsessive-compulsive disorder often have depressive symptoms, and patients suffering from recurrent depressive disorder may develop obsessional thoughts during their episodes of depression. In either situation, increases or decreases in the severity of the depressive symptoms are generally accompanied by parallel changes in the severity of the obsessional symptoms. Obsessive-compulsive disorder is equally common in men and women, and there are often prominent anankastic features in the underlying personality. Onset is usually in childhood or early adult life. The course is variable and more likely to be chronic in the absence of significant depressive symptoms.

- **Diagnostic Guidelines:** For a definite diagnosis, obsessional symptoms or compulsive acts, or both, must be present on mostdays for at least 2 successive weeks and be a source of distress or interference with activities. The obsessional symptoms should have the following characteristics:
    1. They must be recognized as the individual's own thoughts or impulses;
    2. There must be at least one thought or act that is still resisted unsuccessfully, even though others may be present which the sufferer no longer resists;
    3. The thought of carrying out the act must not in itself be pleasurable (simple relief of tension or anxiety is not regarded as pleasure in this sense);
    4. The thoughts, images, or impulses must be unpleasantly repetitive.

Includes:
Anankastic neurosis
Obsessional neurosis
Obsessive-compulsive neurosis

- **Evaluation:** Evaluation is done by history and comparison. Rated using specific anxiety scales like Social anxiety scale by Liebowitz, State and Trait Anxiety Inventory, Hamilton anxiety scale and others (Rose, & Devine, 2014).

- **Treatment:** Mainly there are two modalities of treatment: Pharmacotherapy and Psychotherapy. i). Pharmacotherapy is the use of agents which increases serotonin in the brain like Fluoxetine, serteraline, Mirtazapine, Vellafaxine/ desvellafaxine. Tricyclic anti-depressants to be avoided. ii). Psychotherapy is a form of systematic desensitization of the stimulant. Next is cognitive behavioural therapy, eclectic therapy, meta-cognitive therapy and problem-oriented therapy. Some people show beneficial signs in either or in combination (Pampallona, Bollini, Tibaldi, Kupelnick., & Munizza, 2004).
- **Psychosocial Aspects:** The people who are living with the person who is suffering from the anxiety disorders are mostly found to be ignorant of these things. With the advent of overtly available information in different books or multimedia in their vicinities do know about the problem but are ignorant about the availability of the treatment. The social stigma adds to this. The availability of general hospital psychiatry in the hospitals through the decentralization of psychiatric care availability and perceiving the needs of the people has made a dramatic improvement for the people who are suffering from anxiety or any psychiatric disorders to seek help from the qualified psychiatrists and psychologists available.

PSYCHOTIC DISORDERS

Charaka has documented schizophrenia in his scriptures 33 centuries ago. Morel in 1860 coined the word 'demence precoce'. In 1896, Emil Krapelin coined the term 'dementia precox', which included syndromes catatonia by Kahlbaum (1868). Paranoia by Sander in 1868. Hebephrenia by Hecker in 1870. Schizophrenia was coined by Eugene Bleuler in 1911. He described 4 primary symptoms – Autism, Ambivalence, Loosening of association, Affective flattening. Kurt Schneider described 11 first rank symptoms (Vyas et al., 2011).

1. Acute psychosis
2. Chronic psychosis – Schizophrenia
3. Delusional disorders – Systematized

Non-systematized

For all practical purposes schizophrenia is being taken as the prototype of psychosis. Its prevalence varies across regions in the world, varying from 0.3 per 1000 to 17 per 1000, average is 1.5 to 2.5 per 1000. In high risk population, it is around 2.5 to 3.4 per 1000 population[1.] Incidence in India is 0.93 per 1000 population. Male to female ratio is 1.5:1 in India.

- **Etiopathogenesis:** It is one of the most demonstrable familial diseases. There is 10 times increased risk of schizophrenia in first degree relatives, 50% concordance rate in mono-zygotic twins and 10% concordance in di-zygotic twins. The main impact seen in schizophrenia is over the information processing system (2 types: automatic and serial) involving attention, perception, memory and thinking. McGhie and Chapman showed the failure of sensory filter mechanisms – sensory input processing deficits (ideational gating). Epstein and Kohlman explained cognitive disorganization due to increased sensory input and arousal. Jaynes and Frith implicated a disorder of consciousness, splitting as the basis of thinking disorder in schizophrenia. Magaro explained the inability to integrate perceptional cognitive processes in schizophrenia.

Cognitive Functions

1. **Attention:** In simple terms and best of understanding, it is the ability to focus the mind on a single thing. It is impaired to various extents in schizophrenia.
2. **Consciousness:** Defined as one's own awareness of self and surroundings which includes abstraction also. This is also generally affected.
3. **Language:** Four components- Phonemes, Syntactic, Semantic and Pragmatic. All are affected resulting in loss of prosody and onset of neologisms.
4. **Perception:** It may be over analytical and over detailed at the cost of seeing things in order.
5. **Thought:** Evident in the form of loosening of association, but not specific.
6. **Disorder of Categorization:** Over inclusion and under inclusion affects the conceptual thinking, decision making and problem solving.
7. **Memory:** Affects the immediate memory.

First Rank Symptoms

1. Auditory Hallucinations
    a. First person- designated as I / thought echo
    b. Second person – designated as you
    c. Third person – designated as he/she
2. Thought
    a. Thought insertion
    b. Thought withdrawal
    c. Thought broadcasting
3. Made phenomenon
    a. Made feelings/ affect

   b. Made impulses
   c. Made volitional acts
4. **Somatic Passivity**: Subject is a passive recipient of the bodily sensations imposed by external agency.
5. **Delusional Perception:** Private or illogical meaning attributed to a normal perception.

Evaluation

Done by History taking,
ICD 10 or DSM 5
Positive and negative symptoms of schizophrenia rating scale
Brief psychiatry rating scale

ICD 10

1. Thought echo, thought insertion or withdrawal, and thought broadcasting;
2. Delusions of control, influence, or passivity, clearly referred to body or limb movements
3. Or specific thoughts, actions, or sensations; delusional perception;
4. Hallucinatory voices giving a running commentary on the patient's behaviour, or discussing
5. the patient among themselves, other types of hallucinatory voices coming from some part of the body;
6. Persistent delusions of other kinds that are culturally inappropriate and completely impossible, such as religious or political identity, or superhuman powers and abilities (e.g. being able to control the weather, or being in communication with aliens from another world);
7. Persistent hallucinations in any modality, when accompanied either by fleeting or half-formed delusions without clear affective content, or by persistent over-valued ideas, or when occurring every day for weeks or months on end;
8. Breaks or interpolations in the train of thought, resulting in incoherence or irrelevant speech, or neologisms
9. Catatonic behaviour, such as excitement, posturing, waxy flexibility, negativism, mutism, and stupor;
10. "Negative" symptoms such as marked apathy, paucity of speech, and blunting or incongruity of emotional responses, usually resulting in social withdrawal

and lowering of social performance; it must be clear that these are not due to depression or to neuroleptic medication;

11. significant and consistent change in the overall quality of some aspects of personal behaviour, manifest as loss of interest, aimlessness, idleness, a self-absorbed attitude, and social withdrawal.
    - **Diagnostic Guidelines:** The normal requirement for a diagnosis of schizophrenia is that a minimum of one very clear symptom (and usually two or more if less clear-cut) belonging to any one of the groups listedas (a) to (d) above, or symptoms from at least two of the groups referred to as (e) to (h), should have been clearly present for most of the time during a period of 1 month or more. Conditions meeting such symptomatic requirements but of duration less than 1 month(whether treated or not) should be diagnosed in the first instance as acute schizophrenia-like psychotic disorder and reclassified as schizophrenia if the symptoms persist for longer periods. Symptom (i) in the above list applies only to the diagnosis of Simple Schizophrenia, and duration of at least one year is required. Viewed retrospectively, it may be clear that a prodromal phase in which symptoms and behaviour, such as loss of interest in work, social activities, and personal appearance and hygiene, together with generalized anxiety and mild degrees of depression and preoccupation, preceded the onset of psychotic symptoms by weeks or even months. Because of the difficulty in timing onset, the 1-month duration criterion applies only to the specific symptoms listed above and not to any prodromal nonpsychotic phase. The diagnosis of schizophrenia should not be made in the presence of extensive depressive or manic symptoms unless it is clear that schizophrenic symptoms antedated the affective disturbance. If both schizophrenic and affective symptoms develop together and are evenly balanced, the diagnosis of schizoaffective disorder should be made, even if the schizophrenic symptoms by themselves would have justified the diagnosis of schizophrenia. Schizophrenia should not be diagnosed in the presence of overt brain disease or during states of drug intoxication or withdrawal. Similar disorders developing in the presence of epilepsy or other brain disease, those induced by drugs, should be coded separately (ICD-10, 1993)
    - **Pattern of course:** The course of schizophrenic disorders can be classified by using the following five-Character
    - Continuous,
    - Episodic with progressive deficit,
    - Episodic with stable deficit,
    - Episodic remittent,

- Incomplete remission,
- Complete remission
- Other, Course uncertain, period of observation too short

TREATMENT

Anti-Psychotics

1. **Risperidone** is effective against the positive and negative symptoms
2. **Olanzepine** is found to be effective
3. **Clozapine:** Used for drug resistant schizophrenia. Its use has to be cautious as it can cause agranulocytosis in 1 in 10000 individuals.
4. **Fluoxetine:** May be in need at times, if there is comorbid depression arising as a separate entity or more commonly seen during recovery with the development of insight.

Psychosocial Aspects

- Social skills training
- Cognitive retraining
- Pre-vocational training
- Vocational training
- Job seeking/ Preparation for open employment
    - **Delusional Disorder:** In this disorder the person has a single firm fixed belief which persists and is believed firmly by the person its properties as true. The distinction is that here there is no loss of insight. This does not affect his thought of decision making processes; it is difficult at times to believe the presence of delusion in that person. The insight is completely present. This is of a systemized variety where the delusion builds up in a stepwise manner and a rational explanation can be given. Treatment of them is essential as their presence is not accepted at any time. Treatment of these with antipsychotics will help in creating a wall around the delusion and making them encrypted in some. There are different types: Persecutory type, Jealous type, Erotomanic type, Somatic type [Delusional infestation, Delusional Halitosis, Delusional dysmorphophobia, Miscellaneous disorders], Grandiose type, Shared psychotic disorder[Imposed psychotic disorder, Simultaneous psychotic disorder, Communicated psychotic disorder, Induced psychotic disorder], Acute paranoid disorder[Prison psychosis, Delusional disorders in immigrants and refugees, Culture bound delusional disorders], Late

Figure 2. Flow chart

Paranoid spectrum

Cluster A Personality disorders	Paranoia	Paraphrenia	Paranoid Schizophrenia	Other Schizophrenia	Dementia with Paranoid features

paraphrenia, Paranoid spectrum concept shown in following flow chart, Delusional misidentification syndrome.

SLEEP DISORDERS

Sleep is a condition which is a part of normal physiological process. Sleep has quality and quantity characters. This is guided by circadian rhythm the center of which is present in pons in the hypothalamus. It is an active process. There are many problems associated with sleep and is influenced by many systems in the body like endocrine system, breathing disorders of the respiratory system. Sleep medicine has been recognized as a medical specialty. Normal sleep can be divided into two – non rapid eye movement sleep (NREM sleep) and rapid eye movement sleep (REM sleep). Sleep begins with REM pattern where there is increase in the eye movements present and the individual can have dreams which are remembered most of the times. In EEG Alpha waves are seen during this stage. Each REM stage lasts for 10-20 minutes. The REM latency is shortened in depressive disorders. In this stage, patient can be woken up. NREM sleep is characterized by loss of rapid eye movements and the depth of the sleep increases. Each cycle lasts for 140-180 minutes. Stage 1 lasts for1 to 7 minutes and stage 2 lasts for 10 to 25 minutes, stage 3 for only a few minutes and constitutes about 3 to 8 percent of sleep and stage 4 lasts for 20 to 40 minutes. Dreams occur during the third and fourth stages of sleep. Most of the dreams which occur in the third stage are sometimes recalled. Dreams occurring in the fourth stage, which is the deep sleep stage, with theta (Θ) activity in the EEG and dreams which cannot be recollected. It is told that the thoughts or the actions which are repressed to a sub-conscious level are wishes which could not be fulfilled in the day to day, the apparent fulfillment of which, in turn helps in the homeostasis of the brain functions.

Sleep is affected in many ways. The most common causes are lack of physical activity, increase screen time, life stressors, co morbid psychiatric conditions, substance abuses, last but not the least idiopathic causes. It can be classified into insomnia, sleep related breathing disorders, disorders of hyper somnolence, circardian rhythm sleep-wake disorders, parasomnias, sleep related movement disorders and other sleep disorders.

INSOMNIA

- **Psychophysiological Insomnia**: Conditioned arousal associated with thought of sleep. This occurs with stress and anxiety disorders, delayed sleep phase syndrome and hypnotic drug use and withdrawal. These are usually associated with excessive worry for being unable to sleep, trying hard to sleep, not able to clear the mind when getting into the bed, fall asleep when not trying to, sleeping away from non- bedroom spaces.
- **Idiopathic Insomnia:** It is lifelong inability of the person to obtain adequate sleep. Other causes of insomnia to be ruled out.
- **Paradoxical Insomnia**: The core problem is disassociation between sleep and unconsciousness. The individual thinks he is awake but EEG is consistent with nocturnal sleep.
- **Inadequate Sleep Hygiene:** Insomnia due to behaviors which are non-conducive to good sleep.
- **Behavioral insomnia of childhood:** Children are usually associated with specific behaviors before or during the initiation of sleep. Any modification or changes in the behavior or associated things or persons result in insomnia. If unresolved may develop irregular sleep wake pattern, inadequate sleep hygiene, insufficient sleep syndrome in the later lives.

Insomnia comorbid with mental disorders

Insomnia comorbid with medical condition

Insomnia due to drug or substance use

- **Short Sleeper:** People who require less than five hours of sleep per twenty four hours period in order to maintain day time functioning and mood. It appears they do not have any functional impairment or comorbid conditions. It appears to run in families but specific genes are not identified.
- **Hypersomnolence:** Here the sleep is usually well preserved and efficiency remains very high. The EEG of the sleep is like that of the normal person. Sleep schedules extend for twelve hours or more. Some individuals are having HLA Cw2 positive. Some hypersomnolence are associated with syndromes. Medical conditions associated are Head trauma, stroke, encephalitis, Parkinson's disease, inflammatory conditions, tumors, genetic diseases, neuro-degenerative disorders. Hypersomnolence are sometimes related to menstrual cycle.

- **Narcolepsy**: It is characterized by tetrad of symptoms:
    1. Excessive sleepiness
    2. Cataplexy
    3. Sleep paralysis
    4. Hypnogogic hallucinations.

Hypersomnolence and cataplexy are the key symptoms. Cataplexy is sudden loss of muscle tone, usually with continuing full or partial consciousness; it is commonly precipitated by laughter. It can also be triggered by other strong emotions. The episodes last for seconds to minutes.

Narcolepsy can occur without cataplexy also.

Sleep Related Breathing Disorders

Ranges along several different dimensions involving large airway obstruction during sleep, breathing cessation, central respiratory mechanisms, hypo-ventilation without breathing cessation. It is defined as breathing cessation for 10 seconds or more during sleep.

1. **Obstructive Sleep Apnea/ Hypopnea (OSA):** It may be due to peripheral causes, central causes or mixed type diagnosed as per DSM 5. Comorbidities are hypertension (60%), obesity (30%), Diabetes (33%), angina and MI (28%), Heart failure (14%), stroke including TIA (6%). Psychiatric comorbidity is usually higher that non-sleep apnea group.
2. **Central Sleep Apnea:** Here the cessation of breathing is central in origin due to lack of respiratory effort. Etiologies include heart failure, high altitude, brain stem lesions, metabolic conditions, CNS depressants, congenital abnormalities and positive airwave pressure treatment. Diagnosis based on DSM 5.

Subtypes are

a. Cheyne-stokes breathing
b. Medical disorder without cheyne-stokes breathing
c. High altitude periodic breathing
d. Due to medication or substance
e. Primary central sleep apnea
f. Primary central sleep apnea of infancy
g. Primary central sleep apnea of prematurity.
h. Treatment emergent central sleep apnea

3. **Sleep Related Hypo-Ventilation**: This disorder is associated with chronic obstructive pulmonary disease, obesity and neuro-muscular diseases. Comorbid medical conditions may also contribute and worsen it.

Obesity hypo-ventilation syndrome

a. congenital central alveolar hypo-ventilation syndrome
b. Late onset central hypo-ventilation and hypothalamic dysfunction
c. Idiopathic central alveolar hypo-ventilation
d. Sleep related hypo-ventilation due to medication or substance.
e. Sleep related hypo-ventilation due to medical disorder
f. Sleep related hypoxemia as a separate nosological entity

Diagnosis: Poly-somnography with a) elevated CO2 levels and decreased respiration

If CO2 is not monitored, persistently low levels of oxy-haemoglobin levels not provoked by apnea or hypopnea to be decided by the treating physician

Circadian Rhythm Sleep-Wake Disorder

1. **Delayed Sleep Phase Type 0:** Biological clock runs lower that 24 hours or is shifted later than the desired schedule. They are referred to night-owls
2. **Advanced Sleep Phase Type:** The circadian rhythm cycle is shifted earlier. They are called as early- birds or larks (jokingly called as surgeons in medical centres)
3. **Irregular Sleep Wake Type:** The circadian sleep-wake rhythm is absent or pathologically diminished. They will have normal amount of sleep during the 24 hour period at different times.
4. **Non-24-Hour Sleep-Wake Type:** Here the sleep-wake pacemaker has a cycle length greater or lesser than 24 hours and is not reset each morning. The cycle may be normal for few days, excess or decreased for few days.
5. **Shift Work Type:** Shift workers can suffer from insomnia, excessive sleep or work
6. **Jet-lag:** The natural low point in the normal sleep-wake rhythm occurs approximately at 3-5 am in morning (more prone to accident, industrial and transportation)
7. Unspecified type

Parasomnias and Sleep Related Movement Disorders

Non Rapid Eye Movement Sleep Arousal Disorder (NREMSLAD):

1. **Sleep Walking:** Patient after finishing the REM sleep will go to NREM sleep. This getting up from sleep and doing complex activities during this second or fourth stage of sleep is not remembered by the person. The person returns back and go to sleep from where he had started.

   The following clinical features are essential for a definite diagnosis:
   a. The predominant symptom is one or more episodes of rising from bed, usually during the first third of nocturnal sleep, and walking about;
   b. During an episode, the individual has a blank, staring face, is relatively unresponsive to the efforts of others to influence the event or to communicate with him or her, and can be awakened only with considerable difficulty;
   c. Upon awakening (either from an episode or the next morning), the individual has no recollection of the episode;
   d. Within several minutes of awakening from the episode, there is no impairment f mental activity or behaviour, although there may initially be a short period of some confusion and disorientation;
   e. There is no evidence of an organic mental disorder such as dementia, or a physical disorder such as epilepsy.
2. **Sleep Terrors:** Here it occurs in the stage 4 sleep, non-arousable stage of sleep. The person experiences the dreams which are horrifying and wakes up in panic. The person after being pacified does not get into sleep very easily and the anxiety persists for varied period.

Diagnostic Guidelines

The following clinical features are essential for a definite diagnosis:

1. The predominant symptom is that one or more episodes of awakening from sleep begin with a panicky scream, and are characterized by intense anxiety, body motility, and autonomic hyperactivity, such as tachycardia, rapid breathing, dilated pupils, and sweating;
2. These repeated episodes typically last 1-10 minutes and usually occur during the first third of nocturnal sleep;
3. There is relative unresponsiveness to efforts of others to influence the sleep terror event and such efforts are almost invariably followed by at least several minutes of disorientation and perseverative movements;
4. Recall of the event, if any, is minimal (usually limited to one or two fragmentary mental images);
5. There is no evidence of a physical disorder, such as brain tumour or epilepsy.

- **Nightmare Disorder:** This occurs in the REM stage of the sleep. The horrifying dreams which disturbs the person sleep by waking them up in anxiety is remembered and the person gets pacified after a cup of water consumption and then can go back to sleep after ten to thirty minutes.

Diagnostic Guidelines

The following clinical features are essential for a definite diagnosis:

1. Awakening from nocturnal sleep or naps with detailed and vivid recall of intensely frightening dreams, usually involving threats to survival, security, or self-esteem; the awakening may occur at any time during the sleep period, but typically during the second half;
2. Upon awakening from the frightening dreams, the individual rapidly becomes oriented and alert;
3. The dream experience itself, and the resulting disturbance of sleep, cause marked distress to the individual.

Includes: dream anxiety disorder

Rapid Eye Movement Behavior disorder (REMD):

1. Restless Leg syndrome:
2. Sleep related eating disorders:
3. Recurrent Isolated sleep paralysis
4. Exploring Head Syndrome:
5. Sleep related Hallucinations:
    a. Hypnogogic
    b. Hypnopompic
6. Sleep eneurisis
7. Parasomnia due to a medical disorder
8. Parasomnia drug or substance use
9. Periodic limb movement disorder
10. Sleep related leg cramps
11. Sleep related bruxism
12. Sleep related rhythmic movement disorder
13. Benign sleep myoclonus of infancy
14. Propriospinal myoclonus at sleep onset
15. Sleep related movement disorder due to a medical disorder
16. Sleep related movement disorder due to medications or substance

17. Sleep related movement disorder unspecified

Isolated symptoms and normal variants::

1. Sleep talking
2. Excessive fragmentary myoclonus
3. Hypnogogic Foot Tremor and Alternating Leg Muscle Activation
4. Sleep starts(Hypnic jerks)

Tools in Sleep medicine:

- Clinical interview
- Polysomnography
- Home sleep testing
- Multiple sleep latency test
- Maintaiinence of wakefulness test
- Actigraphy
    - **Treatment:** Treatment of narcolepsy is by using wake promoting substance like modafanil, other drugs are REM sleep suppressing drugs (anti - depressants). Sodium oxibate has proven to be extremely effective in reducing cataplexy. Along with the pharmacological treatment life style modification, psychological counseling and monitoring of general health are also important to be addressed.
    - **Insomnia:** Sedative hypnotic medication, antihistamines, melatonin receptor agonist, orexin antagonist, el-tryptophan are also used.
    - **Non-Pharmacological:** Cognitive behavioral treatment, cognitive therapy, stimulus control therapy, universal sleep hygiene, relaxation therapy and bio-feedback, sleep restriction therapy.

Paradoxical intention: It is a cognitive technique with varying efficacies. The performance anxiety interferes with sleep onset hence the patient tries to stay awake for as long as possible rather than trying to fall asleep.

Obstructive Sleep Apnea (OSA)

1. Positive airwave pressure (PAP
2. Oral appliances
3. Positional therapy
4. Surgical intervention
5. Weight loss

Drug therapy has varied effects

Central sleep apnea – All these can be treated with PAP

Sleep Disorders

This group of disorders includes:

1. **Dyssomnias:** Primarily psychogenic conditions in which the predominant disturbance is in the amount, quality, or timing of sleep due to emotional causes, i.e. insomnia, hypersomnia, and disorder of sleep - wake schedule;
2. **Parasomnias:** Abnormal episodic events occurring during sleep; in childhood these are related mainly to the child's development, while in adulthood they are predominantly psychogenic, i.e. sleepwalking, sleep terrors, and nightmares.

This section includes only those sleep disorders in which emotional causes are considered to be aprimary factor. Sleep disorders of organic origin such as Kleine-Levin syndrome. Nonpsychogenic disorders including narcolepsy and cataplexy and disorders of the sleep - wake schedule, as are sleep apnoea and episodic movement disorders which include nocturnal myoclonus. Finally, enuresis is listed with other emotional and behavioural disorders with onset specific to childhood and adolescence, while primary nocturnal enuresis, which is considered to be due to a maturational delay of bladder control during sleep.

Diagnostic Guidelines

The following are essential clinical features for a definite diagnosis:

1. the complaint is either of difficulty falling asleep or maintaining sleep, or of poor quality of sleep;
2. the sleep disturbance has occurred at least three times per week for at least 1month;
3. there is preoccupation with the sleeplessness and excessive concern over its consequences at night and during the day;
4. The unsatisfactory quantity and/or quality of sleep either causes marked distress or interferes with ordinary activities in daily living.

Treatment

Pharmacological- Benzodiazepines, sedating anti-depressants, lower dose anti-histamines, and low dose melatonin can be used.

Non- pharmacological:-

1. Universal sleep hygiene
2. Stimulus control therapy
3. Sleep restriction therapy
4. Cognitive therapy
5. Cognitive behavioral therapy
6. Paradoxical intention therapy

In this the person tries to stay awake as long as possible rather than trying to fall asleep. Here the performance anxiety will be reduced and sleep latency will be improved.

Treatment of circadian – rhythm sleep- wake disorder:

1. Light Therapy
2. Medications
3. Chronotherapy

CHILDHOOD DISORDERS

1. Intellectual Disability Disorders (IDD) or Mental Retardation (MR)
2. Developmental learning disorders
3. Childhood psychiatric disorders
4. Autism spectrum disorders
5. Attention Deficit/ Hyperactivity disorders
6. Other disorders like Nocturnal Enuresis
7. Tic disorders
8. Eating disorders

Further childhood disorders explained in following order

IDD

IDD is defined as the discrepancy in the growth of body and mind. Here, there is body/ physical growth according to the age. But, the mind or the intelligence does not match with that of the growing physical age. The mind grows till the age of 16yrs. Average intelligence ranges between 90-110. IQ less than 90 is defined as lower intelligence. IQ more than 110 is designated to be more intelligent. Hence the intelligence (IQ) is being calculated as below:

IQ= Mental age × 100

Chronological age

The prevalence of mental retardation (MR) varies from place to place. It ranges from 2-6% per 1000 children. Intelligence between 71-90 is designated to be borderline mental retardation. IQ of 51-70 are designated as mild mental retardation. IQ 0f 36-50 is defined as moderate mental retardation. IQ between 21-35 is designated as severe mental retardation. IQ less than 20 is designated as profound mental retardation. Usually there would be association of neurological problems in severe and profound mental retardation.

Etiopathogenesis:

The causes of mental retardation could be divided into

1. **Primary:** Mental retardation due to biological causes (Genetic inheritance) which is seen since birth
2. **Secondary:** This occurs after a period of normal mental growth. Causes being meningitis, cerebritis, head trauma – subdural hematoma, extradural hematoma; space occupying lesions
    a. **Prenatal:** Genetic load, Prenatal infections in the mother like TORCH infections, Metabolic syndromes like phenylketonuria, storage disorders, endocrine disorders
    b. **Natal:** hypoxic cerebral injuries, infections, birth trauma
    c. **Postnatal:** CNS infections, trauma, endocrine disorders,
        - **Clinical Features:** In Primary or prenatal/natal causes, the milestones, language, motor, comprehension, skills may be delayed. In secondary causes, till the insult is been present, the growth would be normal and is delayed from the onset of trauma. Mild MR children are able to do some physical activities on supervision. Moderate MR Persons can be trained to maintain their routine works on supervision only. A severe and profound MR child has to be completely cared by the caregivers. Borderline MR Children needs separate attention in teaching and can attain formal education with difficulty. Around 10% of people with Cerebral palsy are usually associated in patients with mental retardation. 25% of the children with IDD have comorbid epilepsy.
        - **Diagnosis:** Using ICD 10/ DSM 5

Evaluation

IDD is being evaluated using

1. Seguin form board
2. Binet Kamath Test of Intelligence
3. Weschler's Intelligence scale
4. Raveen's progressive matrices
5. Bhatia's battery
    - **Treatment:** There are no any forms of treating the cause and rectifying the sequelae happened, as the neurons once damaged cannot regenerate; hence their functions. The only way is to re-train them. This is a multi-disciplinary approach--behavioral problems, Seizures, Psychotic features can be managed by the drugs. Cerebral Palsy needs Physiotherapy and rarely surgical interventions.
    - **Psychosocial Aspect:** This disorder is a great unexpected part by the caregivers as it is usual that they will always have positive attitude towards their child that they will grow to be the better off than them or as other child. During the process of acceptance of the Mental retardation they may have many Psychological reactions amounting to Anxiety, Depression or Psychotic reactions also. It may be transient or pervasive also. The work of the Psychiatrist, Psychologist and Psychiatric Social Workers are to prevent it, make them accept the reality and coordinate with the health staff in the process of retraining/ rehabilitation works.
    - **Developmental Learning Disorders:** In this type of problems the child is unable to learn certain things. It may vary from language, behavior, specific subject. It may or may not be associated with IDD. These children are benefitted from Government by way of exemption in the exams.
    - **Childhood Psychiatric Disorders:** All forms of psychiatric illness which is seen in adults can occur in the children also with mild variations in its presentations. Treatment here should be with empathy, Pharmacological and non-pharmacological interventions are at most important.
    - **Autism Spectrum Disorders:** This occurs in children who are having and attained normal development for some time. It usually begin by the age of 2-3 years and is characterized by loss of learned behavior and not responding to the usually stimuli. The characteristic feature is that the child sees through the eyes of the person in front of him and has his own way of interpreting the meaning of the situation.

- **Attention Deficit/ Hyperactivity Disorders:** This is one of the most common disorder of the childhood and at times can be seen in the adulthood also. The prevalence varies and varies with the age group. Its incidence in Mumbai is 8.7% in children going to primary school. The Male to Female ration is 7.6:1. This has multiple comorbidities along with it. Genetic component of inheritance is also present. The main factor is short attention span. It needs a holistic approach and other comorbidity needs to be ruled out.
- **Interventions:** Parental counseling, Behavioural interventions, Pharmacotherapy, Psychotherapeutic interventions and Education.
- **Nocturnal Enuresis:** It is defined as passage of urine in the night time during sleep without his/her knowledge, after the age of 5 years. It may be primary or secondary. There are multiple causes. The organic factors as the cause to be ruled out first and then the psychiatric cause to be searched for. Treatment consists of pharmacological, behavior therapy, retraining.
- **Tic Disorders:** It is defined as involuntary movement of the muscle fibres in toto or individually. It gets represented as twitching, vocal sounds. It is of two types, Motor tics, Vocal tics

EATING DISORDERS

Anorexia Nervosa

Anorexia nervosa is a disorder characterized by deliberate weight loss, induced and/or sustained by the patient. The disorder occurs most commonly in adolescent girls and young women, but adolescent boys and young men may be affected more rarely, as may children approaching puberty and older women up to the menopause. Anorexia nervosa constitutes an independent syndrome in the following sense:

1. The clinical features of the syndrome are easily recognized, so that diagnosis is reliable with a high level of agreement between clinicians;
2. Follow-up studies have shown that, among patients who do not recover, a considerable number continue to show the same main features of anorexia nervosa, in a chronic form.
3. Body weight is maintained at least 15% below that expected (either lost or never achieved), or Quetelet's body-mass index4 is 17.5 or less.4 Quetelet's

4. Body-mass index = weight (kg) to be used for age 16 or more. Prepubertal patients may show failure to make the expected weight gain during the period of growth.
5. The weight loss is self-induced by avoidance of "fattening foods". One or more of the following may also be present: self-induced vomiting; self-induced purging; excessive exercise; use of appetite suppressants and/or diuretics.
6. There is body-image distortion in the form of a specific psychopathology whereby a dread of fatness persists as an intrusive, overvalued idea and the patient imposes a low weight threshold on himself or herself.
7. A widespread endocrine disorder involving the hypothalamic - pituitary -gonadal axis is manifest in women as amenorrhoea and in men as a loss of sexual interest and potency. (An apparent exception is the persistence of vaginal bleeds in anorexic women who are receiving replacement hormonal therapy, most commonly taken as a contraceptive pill.) There may also be elevated levels of growth hormone, raised levels of cortisol, changes in the peripheral metabolism of the thyroid hormone, and abnormalities of insulin secretion.
8. If onset is pre-pubertal, the sequence of pubertal events is delayed or even arrested (growth ceases; in girls the breasts do not develop and there is a primary amenorrhoea; in boys the genitals remain juvenile). With recovery, puberty is often completed normally, but the menarche is late.
    - **Bulimia Nervosa:** is a syndrome characterized by repeated bouts of overeating and an excessive preoccupation with the control of body weight, leading the patient to adopt extreme measures so as to mitigate the "fattening" effects of ingested food. Anorexia nervosa by virtue of sharing the same psychopathology. The age and sex distribution is similar to that of anorexia nervosa, but the age of presentation tends to be slightly later. The disorder may be viewed as a sequel to persistent anorexia nervosa (although the reverse sequence may also occur). A previously anorexic patient may first appear to improve as a result of weight gain and possibly a return of menstruation, but a pernicious pattern of overeating and vomiting then becomes established. Repeated vomiting is likely to give rise to disturbances of body electrolytes, physical complications (tetany, epileptic seizures, cardiac arrhythmias, muscular weakness), and further severe loss of weight.

Diagnostic Guidelines

For a definite diagnosis, all the following are required:

1. There is a persistent preoccupation with eating, and an irresistible craving for food; the patient succumbs to episodes of overeating in which large amounts of food are consumed in short periods of time.
2. The patient attempts to counteract the "fattening" effects of food by one or more of the following: self-induced vomiting; purgative abuse, alternating periods of starvation; use of drugs such as appetite suppressants, thyroid preparations or diuretics. When bulimia occurs in diabetic patients they may choose to neglect their insulin treatment.
3. The psychopathology consists of a morbid dread of fatness and the patient sets herself or himself a sharply defined weight threshold, well below the premorbid weight that constitutes the optimum or healthy weight in the opinion of the physician. There is often, but not always, a history of an earlier episode of anorexia nervosa, the interval between the two disorders ranging from a few months to several years. This earlier episode may have been fully expressed, or may have assumed a minor cryptic form with a moderate loss of weight and/or a transient phase of amenorrhoea.

PERSONALITY DISORDERS (PD)

Personality as defined by Theodore Millon is of ingrained, pervasive, enduring, and habitual ways of psychological function. This is unique to each individual. It is inter-related organization of attitudes, perceptions, habits, emotions and behaviors. Traits are defined as enduring patterns of perceiving, relating, in thinking about the surroundings and oneself which are important in the social and personal context. Personality involves predispositions/traits, mind processes (synthesizing, integrating, recognition of motivation) when these traits or behavior are inflexible and maladaptive cause significant functional impairment and subjective distress. Temperaments are biologically based predispositions, 4 bodily humors (melancholic, irritable, optimistic and apathetic) which colors the personality. Genetically based individual variation emerges early in life, remains relatively stable with increasing age. James Pritchard, father of personality disorders (moral insanity) defined them as morbid perversion of natural feelings, afflictions, inclinations, temper, habits, moral disposition and natural impulses without any marked disorder of intellect, knowing and reasoning. Features of personality disorders are

1. Early onset – Adolescence or early adulthood
2. Stable and persistent
3. Pervasive not restricted to any one kind of situation and has an effect on variety of domains – family.

4. Interpersonal focus
5. Impairment – it crosses from abnormality to disorder

To know about the personality disorder as a background of people committing crimes, behaving a socially and not following the social norms and the legal laws framed by the constitution leading to crimes and disruption of morality of individual, interpersonal and social walks of life. This also has increased mortality in the form of suicide, accidents, and homicide. These persons may have psychotic or neurotic psychiatric disorders comorbid.

Classification

Phillippe Pinel in 1809 started to define and classify. Pritchard – moral insanity in 1835. Freud in 1908 defined hysterical personality; Carl Abraham in 1949 defined dependent obsessive compulsive and hysterical personality disorders. Emile Kraeplin stated the concept of spectrum of personality disorders. Kurt Schneider in 1937 and 1934 used social deviance as a marker for personality disorders. Tierer and Seivewright in 1998 described two groups of personality disorders 1. Immature ones, that resolves over time and 2. Mature ones that tend to persist well into adult life.

- **Diagnostic Guidelines:** Conditions not directly attributable to gross brain damage or disease, or to another psychiatric disorder, meeting the following criteria:
    1. Markedly disharmonious attitudes and behaviour, involving usually several areas of functioning, e.g. affectivity, arousal, impulse control, ways of perceiving and thinking, and style of relating to others;
    2. The abnormal behaviour pattern is enduring, of long standing, and not limited to episodes of mental illness;
    3. The abnormal behaviour pattern is pervasive and clearly maladaptive to abroad range of personal and social situations;
    4. The above manifestations always appear during childhood or adolescence and continue into adulthood;
    5. The disorder leads to considerable personal distress but this may only become apparent late in its course;
    6. The disorder is usually, but not invariably, associated with significant problems in occupational and social performance.

For different cultures it may be necessary to develop specific sets of criteria with regard to social norms, rules and obligations. For diagnosing most of the subtypes listed below, clear evidence is usually required of the presence of at least three of the traits or behaviours given in the clinical description.

Mature Personality Disorders

1. Anankastic PD (obsessive compulsive)
2. Paranoid
3. Schizoid (unschizotypal)
4. Anxious (avoidant)

Immature Personality Disorders

1. Antisocial)Dissocial)
2. Borderline (Impulsive)
3. Histrionic
4. Dependent
5. Narcissistic

In 1979, WHO developed a international personality disorder examination (IPDE) which examines under six headings like

1. Work
2. Self
3. Interpersonal relationships,
4. Affect
5. Reality testing
6. Impulse – control

It has undergone modifications over the years. In personality disorders the prevalence is found to be 6-11% in older studies and newer reports show around 4-8%

- **Paranoid Personality Disorder:** Personality disorder characterized by;
    1. Excessive sensitiveness to setbacks and rebuffs;
    2. Tendency to bear grudges persistently, e.g. refusal to forgive insults and injuries or slights;
    3. Suspiciousness and a pervasive tendency to distort experience by misconstruing the neutral or friendly actions of others as hostile or contemptuous;

4. A combative and tenacious sense of personal rights out of keeping with the actual situation;
5. Recurrent suspicions, without justification, regarding sexual fidelity of spouse or sexual partner;
6. Tendency to experience excessive self-importance, manifest in a persistent self-referential attitude;
7. Preoccupation with unsubstantiated "conspiratorial" explanations of events both immediate to the patient and in the world at large.

Includes: expansive paranoid, fanatic, querulant and sensitive paranoid personality disorder

- **Schizoid Personality Disorder:** Personality disorder meeting the following description;
    1. Few, if any, activities, provide pleasure;
    2. Emotional coldness, detachment or flattened affectivity;
    3. Limited capacity to express either warm, tender feelings or anger towards others;
    4. Apparent indifference to either praise or criticism;
    5. Little interest in having sexual experiences with another person (taking into account age);
    6. Almost invariable preference for solitary activities;
    7. Excessive preoccupation with fantasy and introspection;
    8. Lack of close friends or confiding relationships (or having only one) and of desire for such relationships;
    9. Marked insensitivity to prevailing social norms and conventions.
- **Dissocial Personality Disorder:** usually coming to attention because of a gross disparity between behaviour and the prevailing social norms, and characterized by:
    1. Callous unconcern for the feelings of others;
    2. Gross and persistent attitude of irresponsibility and disregard for social norms, rules and obligations;
    3. Incapacity to maintain enduring relationships, though having no difficulty in establishing them;
    4. Very low tolerance to frustration and a low threshold for discharge of aggression, including violence;
    5. Incapacity to experience guilt or to profit from experience, particularly punishment;
    6. Marked proneness to blame others, or to offer plausible rationalizations, for the behaviour that has brought the patient into conflict with society.

There may also be persistent irritability as an associated feature. Conduct disorder during childhood and adolescence, though not invariably present, may further support the diagnosis.

Includes: amoral, antisocial, asocial, psychopathic, and sociopathic personality (disorder)

- **Borderline Personality Disorder:** More patients fit borderline personality disorder also meet the criteria for histrionic PD. Psychoanalysts describe a group of patients with neurotic disorders who are prone to brief psychotic episodes under stress including the stress of psycho-analysis. These people are intolerant about loneliness and abandonment and have great difficulty dealing with these issues. They respond better to SSRI's. Psychotherapy along with pharmacotherapy are helpful in some. Several of the characteristics of emotional instability are present; in addition, the patient's own self-image, aims, and internal preferences (including sexual) are often unclear or disturbed. There are usually chronic feelings of emptiness. A liability to become involved in intense and unstable relationships may cause repeated emotional crises and may be associated with excessive efforts to avoid abandonment and a series of suicidal threats or acts of self-harm (although these may occur without obvious precipitants).

Includes: borderline personality (disorder) (APA, 2013).

- **Impulsive Type Personality Disorder:** The predominant characteristics are emotional instability and lack of impulse control. Outbursts of violence or threatening behaviour are common, particularly in response to criticism by others.

Includes: explosive and aggressive personality (disorder)

- **Antisocial Personality Disorder (ASPD):** The people tend to behave or act in a manner that violates the social norms. In US, the prevalence in males was around 4.5% and in females 0.8%/ Studies in prison have shown the prevalence of ASPD between 39 – 76%. In genetics, criminality has been equated with ASPD. Concordance of criminality in twins in mono-zygotic twins was 51% and a di-zygotic twin was 22%, suggestive of a definite genetic component. In juvenile delinquency, there was little difference between monozygotic 87% and dizygotic 72% twins suggesting familial component. There is a high prevalence of ASPD among relatives of women diagnosed as having Briquette's Syndrome.

- **Schizotypal Personality Disorder:** Odd, eccentric group is characterized by social detachment and idiosyncratic behavior. This includes schizotypal, paranoid and schizoid PD. They are common in phenomenological, genetic, biological, outcome and treatment response characteristics studies.

Anxious Avoidant Personality Disorder

They exhibit social withdrawal, poor self-esteem and social anxiety.

Personality disorder characterized by;

1. Persistent and pervasive feelings of tension and apprehension;
2. Belief that one is socially inept, personally unappealing, or inferior to others;
3. Excessive preoccupation with being criticized or rejected in social situations;
4. Unwillingness to become involved with people unless certain of being liked;
5. Restrictions in lifestyle because of need to have physical security;
6. Avoidance of social or occupational activities that involve significant interpersonal contact because of fear of criticism, disapproval, or rejection. Associated features may include hypersensitivity to rejection and criticism.
    - **Anankastic Personality Disorder:** It has a high genetic loading showing OCD proband-wise concordance of 13 out of 18 in mono-zygotic twins and 7 out of 15 in dizygotic twins

Personality disorder characterized by:

1. Feelings of excessive doubt and caution;
2. Preoccupation with details, rules, lists, order, organization or schedule;
3. Perfectionism that interferes with task completion;
4. Excessive conscientiousness, scrupulousness, and undue preoccupation with productivity to the exclusion of pleasure and interpersonal relationships;
5. Excessive pedantry and adherence to social conventions;
6. Rigidity and stubbornness;
7. Unreasonable insistence by the patient that others submit to exactly hisor her way of doing things, or unreasonable reluctance to allow others to do things;
8. Intrusion of insistent and unwelcome thoughts or impulses.
    - **Includes:** Compulsive and obsessional personality (disorder)
    - **Histrionic Personality Disorder:** This can be classified under somatization or conversion disorder. In the later classification of DSM 4. Histrionic PD was named. Features are emotional instability/aggression, over-involvement/emotionality, dependent and exhibitionistic/oral aggressions, pseudo hypersexuality/exhibitionism, sexual inhibition/ egocentricity, competitiveness/sexual proactiveness, masochism/ dependency, obstinacy.

Personality disorder characterized by:

1. Self-dramatization, theatricality, exaggerated expression of emotions;
2. Suggestibility, easily influenced by others or by circumstances;
3. Shallow and labile affectivity;
4. Continual seeking for excitement and activities in which the patient is the centre of attention;
5. Inappropriate seductiveness in appearance or behaviour;
6. Over-concern with physical attractiveness.

Associated features may include egocentricity, self-indulgence, continuous longing for appreciation, feelings that are easily hurt, and persistent manipulative behaviour to achieve own needs.

Includes: hysterical and psychoinfantile personality (disorder)

- **Emotionally Unstable Personality Disorder:** A personality disorder in which there is a marked tendency to act impulsively without consideration of the consequences, together with affective instability. The ability to plan ahead may be minimal, and outbursts of intense anger may often lead to violence or "behavioural explosions"; these are easily precipitated when impulsive acts are criticized or thwarted by others. Two variants of this personality disorder are specified, and both share this general theme of impulsiveness and lack of self-control.

Dependent Personality Disorder

Personality disorder characterized by:

1. Encouraging or allowing others to make most of one's important life decisions;
2. Subordination of one's own needs to those of others on whom one is dependent, and undue compliance with their wishes;
3. Unwillingness to make even reasonable demands on the people one depends on;
4. Feeling uncomfortable or helpless when alone, because of exaggerated fears of inability to care for oneself;
5. Preoccupation with fears of being abandoned by a person with whom one has a close relationship, and of being left to care for oneself;
6. Limited capacity to make everyday decisions without an excessive amount of advice and reassurance from others.

Associated features may include perceiving oneself as helpless, incompetent, and lacking stamina.

Includes: asthenic, inadequate, passive, and self-defeating personality (disorder)

- **Other Specific Personality Disorders:** A personality disorder that fits none of the specific rubrics

Includes: eccentric, "haltlose" type, immature, narcissistic, passive aggressive, and psychoneurotic personality (disorder)

- **Management:** All PD's need pharmacotherapy and psychotherapy at varying levels for its behavioral modification. The traits or behaviors core to the PD's tend not to change with any modalities, hence therapy is individualized based on the extent of trait amplification in the form of behavior in that person.
- **Sexual Disorders:** Sexual dysfunctions are disturbances in – sexual desire, changes in sexual response cycle leading to marked distress and interpersonal difficulty.
    1. Sexual desire disorders
        a. Hypoactive sexual desire
        b. Sexual aversion disorder
    2. Sexual arousal disorder
        a. Female sexual arousal disorder
        b. Male sexual arousal disorder
    3. Orgasmic disorders
        a. Female orgasmic disorder
        b. Male orgasmic disorder
        c. Premature ejaculation
    4. Sexual pain disorder
        a. Dyspareunia
        b. Vaginismus
    5. Sexual Dysfunction due to general medical conditions
    6. Substance induced sexual dysfunction
    7. Sexual dysfunction, NOS
    8. Paraphillias

It is Characterized by recurrent, intense sexual urges, fantasies or behaviors that involve unusual objects, activities or situations and cause clinically significant distress in social, occupational and other important areas of functioning. It includes

a. Exhibitionism

b. Fetishism
c. Frotteurism
d. Peadophilia
e. Sexual masochism
f. Sexual sadism
g. Transvestic fetishism
h. Voyeurism
i. Paraphilia, NOS

- **Sexual Disorders, NOS:** The sexual dysfunctions maybe due to biological or psychological factors or both. The sexual dysfunction maybe lifelong or acquired; it maybe generalized or situational. The etiological factor varies from social causes, vascular causes, drugs, endocrine disorders, psychological causes in many of the sexual dysfunctions.
- **Management:** Specific drugs for erectile dysfunctions like sildenafil. Mechanical devices, surgical treatment in rare cases. The use of foreplay before the sexual act determines the symptoms for many sexual dysfunctions. At times, behavioral therapy may be beneficial in some.

REFERENCES

American Psychiatric Association. (2013). *Diagnostic and Statistical Manual of Mental Disorders* (5th ed.). Arlington, VA: American Psychiatric Association.

Graziane & Sweet. (2017). Comprehensive textbook of psychiatry (10th ed.). Wolters Kluwer.

Jellinger, K. A. (2010). Should the word 'dementia' be forgotten? *Journal of Cellular and Molecular Medicine*, *14*(10), 2415–2416. doi:10.1111/j.1582-4934.2010.01159.x PMID:20731751

Kotzbauer, P. T., Trojanowsk, J. Q., & Lee, V. M. (2001). Lewy body pathology in Alzheimer's disease. *Journal of Molecule Neuroscience, 17*(2), 225-32.

Liz, S. (2016). *Dementia Rates Decline Sharply Among Senior Citizens*. Retrieved from https://khn.org/news/dementia-rates-decline-sharply-among-senior-citizens/

Nasreddine, Z. S., Phillips, N. A., Bédirian, V., Charbonneau, S., Whitehead, V., Collin, I., & Chertkow, H. (2005). The Montreal Cognitive Assessment, MoCA: A brief screening tool for mild cognitive impairment. *Journal of the American Geriatrics Society*, *53*(4), 695–699. doi:10.1111/j.1532-5415.2005.53221.x PMID:15817019

Pampallona, S., Bollini, P., Tibaldi, G., Kupelnick, B., & Munizza, C. (2004). Combined pharmacotherapy and psychological treatment for depression: a systematic review. *Archives of General Psychiatry, 61*(7), 714-9.

Reinhard, S. C., Given, B., Petlick, N. H., & Bemis, A. (2008). Supporting Family Caregivers in Providing Care. In *Patient Safety and Quality: An Evidence-Based Handbook for Nurses*. Rockville, MD: Agency for Healthcare Research and Quality. Available from: https://www.ncbi.nlm.nih.gov/books/NBK2665/

Rose, M., & Devine, J. (2014). Assessment of patient-reported symptoms of anxiety. *Dialogues in Clinical Neuroscience, 16*(2), 197–211. PMID:25152658

Sadock, B. J., Kaplan, H. I., & Sadock, V. A. (2005). *Kaplan & Sadock's Comprehensive Textbook of Psychiatry*. Philadelphia: Lippincott Williams & Wilkins.

The ICD-10. (1993). *Classification of Mental and Behavioural Disorders, Clinical descriptions and diagnostic guidelines*. World Health Organization.

Vyas, J., & Ahuja, N. (2011). Textbook of Postgraduate Psychiatry (2nd ed..). Jaypee Brothers, Medical Publishers (p) Ltd.

Whooley, M. A., Grady, D., & Cauley, J. A. (2000). Postmenopausal estrogen therapy and depressive symptoms in older women. *Journal of General Internal Medicine, 15*(8), 535–541. doi:10.1046/j.1525-1497.2000.04029.x PMID:10940144

KEY TERMS AND DEFINITIONS

Acetylcholine: (ACh) is an organic chemical that functions in the brain and body of many types of animals (and humans) as a neurotransmitter—a chemical message released by nerve cells to send signals to other cells, such as neurons, muscle cells, and gland cells.

Anxiety: Anxiety is your body's natural response to stress. It's a feeling of fear or apprehension about what's to come. The first day of school, going to a job interview, or giving a speech may cause most people to feel fearful and nervous

Dementia: Is a syndrome in which there is deterioration in memory, thinking, behaviour, and the ability to perform everyday activities. Although *dementia* mainly affects older people, it is not a normal part of ageing

Depression: Depression is the main cause of disability worldwide, according to the World Health Organization (WHO). It can affect adults, adolescents, and children.

Dopamine: (DA, a contraction of 3,4-dihydroxyphenethylamine) is an organic chemical of the catecholamine and phenethylamine families. It functions both as a hormone and a neurotransmitter, and plays several important roles in the brain and body.

Mania: Also known as *manic syndrome*, is a state of abnormally elevated arousal, affect, and energy level, or a state of heightened overall activation with enhanced affective expression together with lability of affect. Although mania is often conceived as a "mirror image" to depression, the heightened mood can be either euphoric or irritable; indeed, as the mania intensifies, irritability can be more pronounced and result in violence, or anxiety.

Obsessive-Compulsive Disorder (OCD): Is a mental illness that causes repeated unwanted thoughts or sensations (obsessions) or the urge to do something over and over again (compulsions). Some people can have both obsessions and compulsions.

Psychosis: Is an abnormal condition of the mind that results in difficulties determining what is real and what is not. Symptoms may include false beliefs (delusions) and seeing or hearing things that others do not see or hear (hallucinations).

Serotonin: Is a chemical nerve cells produce. It sends signals between your nerve cells. *Serotonin* is found mostly in the digestive system, although it's also in blood platelets and throughout the central nervous system. *Serotonin* is made from the essential amino acid tryptophan.

Substance Use: The term "substance use" refers to the use of drugs or alcohol, and includes substances such as cigarettes, illegal drugs, prescription drugs, inhalants and solvents. A substance use problem occurs when using alcohol or other drugs causes harm to you or to others. Substance use problems can lead to addiction.

Chapter 11
Functional Human Brain Mapping

Shahzadi Malhotra
Chacha Nehru Hospital, India

Gaurav Rajender
SMS Medical College and Hospital, India

ABSTRACT

Neuroimaging plays a crucial role in psychiatry and mental health as it can potentially be used to identify causal factors, biomarkers of a disorder, prognosis or treatment, elucidate biological pathways along with informing and monitoring newer therapies. Early and prophylactic interventions present an emerging future direction in clinical psychiatry, and neuroimaging has the potential to help in identification of individuals at risk and monitor the effects of intervention. In the chapter an attempt is made to elaborate the concept of brain mapping, and differences between structural and functional brain imaging are discussed. The chapter focusses on advent of neuroimaging in psychiatry along with discussing the major functional neuroimaging techniques.

CHAPTER OUTLINE

- Introduction
- History of neuroimaging and brain mapping
- Difference between structural and functional brain mapping
- Importance of brain mapping in psychiatric disorders
- Methods of study in brain mapping

DOI: 10.4018/978-1-7998-2860-0.ch011

learning objectives

- To describe human brain mapping
- To describe the historical advent of brain mapping
- To differentiate between structural and functional brain mapping
- To describe the importance of brain mapping in neuropsychiatric disorders
- To describe the various methods of brain mapping

INTRODUCTION

Neuroimaging or brain imaging is the use of various techniques to either directly or indirectly image the structure, function, or pharmacology of the nervous system. It is a relatively new discipline within medicine, neuroscience and psychology. Neuroimaging is the discipline that deals with the *in vivo* depiction of anatomy and function of the central nervous system (CNS) in health and disease. The term 'neuroimaging' is more comprehensive than the older term of neuroradiology; although in the past, neuroimaging was considered the domain of radiologists with a specific interest in the nervous system, the participants in this rapidly changing field now come from diverse backgrounds. Contributors exist in all areas in the neurosciences, including basic scientists, neurologists, neurosurgeons, psychologists, psychiatrists, physicists, chemists, and clinicians in radiology and nuclear medicine. It is important to realize that neuroimaging is a dynamic and evolving field.

In many clinical situations, neuroimaging plays the most decisive role among neurobiological investigations. Over the past four decades, the field of cognitive neuroscience has emerged as an important area in neurosciences. Cognitive neuroscience combines the experimental strategies of cognitive psychology with various techniques to examine how brain function supports mental activities (Raichelle, 2008). Neurobiological factors are important determinants of brain functioning and psychiatric illnesses. However, the cause and effect relationship between these neurobiological factors and observed psychiatric symptoms is still not well established. Although, both direct (CT scan, MRI, electrophysiological recordings) and indirect methods (plasma, urine, cerebrospinal fluid studies have established the association of neurobiological factors with psychiatric symptoms; however, these findings and associations are still under evaluation. There are many neuroimaging techniques that have been developed over years and have helped our understanding of neurobiological basis of behavior. The most commonly used neuroimaging techniques that interest students of clinical psychology will be discussed in this chapter.

ADVENT OF NEUROIMAGING IN PSYCHIATRY

The history of neuroimaging techniques in Psychiatry dates back to 1976, when Johnstone demonstrated that patients with schizophrenia have enlarged cerebral ventricles (Johnstone, 1976). At that time, the major emphasis was that mental disorders were largely caused due to early life experiences along with some biological influences. In the last 40 years, modern psychiatry has come out of this dualistic approach to etiology of mental illnesses, and there is now general agreement among mental health experts that environmental influences can manifest themselves in observable brain changes as well as genetic factors (Linden, 2012). The biggest advantage of this newer understanding of mental illnesses in terms of a various etiological factors combining biological as well as environmental approaches is the emergence of research programmes that aim to map out neural correlates of mental illness and also map the predictors of psychotherapy by using functional neuroimaging (Beutel, et al., 2003; DeRubeis et al., 2008; Linden, 2006, 2008, 2012). Thus, neuroimaging plays an important role in establishing the biological foundations of psychiatric diagnosis, as neuroimaging techniques are the only way to study the living human brain.

STRUCTURAL VS FUNCTIONAL BRAIN MAPPING

Brain mapping can take many forms, but brain maps of all types can be generally divided as structural maps or functional maps. Structural mapping reveals and gives an understanding of the neural hardware, that is the anatomical connectivity between one part of the brain to another and the principles that apply to the way in which these linkages are ordered and placed. Methos for structural mapping of the brain almost always consist of electrical stimulation of neurons and neuronal pathways. The measured output variable may take several forms, including intracellular potential, field potentials reflecting underlying synaptic inputs to an area of the brain, extracellularly recorded unit potentials, or electromyographic (EMG) activity. The main distinguishing factor in structural mapping is that a part of the system is stimulated either magnetically or electrically, and a response, usually in the form of some type of electrical activity is recorded from another part of the neural circuit.

Unlike structural methods of brain mapping, functional mapping as the name itself indicates, reveals the spatial representation within the brain of some natural parameter such as touch receptors on the body surface, movement of a body part, the frequency of an auditory stimulus, or the local of a light stimulus in the visual field.. Functional mapping involves the orderly localization of natural stimuli or movements to examine organization within a particular brain region. An important principle underlying all functional imaging techniques is that cerebral function is closely related to cerebral metabolism, which in turn is couples to blood flow (Sawle, 1995). While considering activation studies, in which changes in brain function associated with some specific task or sensory stimulation are studied, an important principle is that when the brain is engaged in any activity, there are focal, or regional increase in blood flow in that part of the brain which is specifically engaged in that task, and that this blood flow is disproportionately large as compared with any associated general increase in cerebral blood flow.

IMPORTANCE OF BRAIN MAPPING IN NEUROPSYCHIATRIC DISORDERS

Psychiatric disorders involve abnormalities in the thoughts, behaviours and mood. Understanding the process and causes underlying these conditions has progressed from the notions that an imbalance of black bile caused melancholia to complex formulations of interacting genetic, social, psychological and environmental factors. (Gelder, 1996; Fu & McGuire, 1999). Unlike many physical and neurological disorders, psychiatric disorders do not cause changes visible to the naked eye in the neuroimaging study of the individual patient (Agarwal, 2010). Although, the manifestations of psychiatric illnesses are difficult to attribute to focal brain lesions, these psychiatric disorders reflect an underlying brain dysfunction; which may be caused by either social, psychological, environmental, biological or genetic factor. The emergence of neuroimaging has provided an unprecedented opportunity to relate cognitive and behavioural abnormalities to changes in brain function (Fu & McGuire, 1999). Psychiatric disorders can now be studied and investigated by various neuroimaging modalities, particularly quantitative structural imaging, as exampled by voxel-based morphometry, and functional neuroimaging, using magnetic resonance imaging (MRI) techniques, positron emission tomography (PET), and single photon emission tomography (SPECT).

In many instances, neuroimaging may help the clinician to reach at the correct diagnosis in a patient presenting with psychiatric symptoms which may many a times be caused by neurological diseases masking as psychiatric disorders or by disorders currently considered to be primarily psychiatric in nature (Masdeu, 2011). Neuroimaging is routinely used in the workup of patients with psychotic disorders because lesions of the frontal or temporal lobes, most often tumors, can present with psychosis (Bunevicious et al. 2008). In some cases, a behavioral syndrome is caused by focal seizures arising from a tumor, or from congenital or traumatic lesions (Sherwin et al., 1982, Boone et al., 1988). Psychotic features may also be found with thalamic or hypothalamic lesions (Moise et al., 2006; Albert et al., 1988). Apathy caused by a frontal brain tumor may be mistaken for depression (Grafman et al., 1986). In older people with cognitive impairment, it may be difficult to differentiate a neurodegenerative disorder from depression. Neuroimaging may be helpful in these situations by showing findings characteristic of Alzheimer's disease (AD), diffuse Lewy body disease, or one of the frontotemporal dementias. AD is characterized by medial temporal atrophy on MRI or X-ray computed tomography (CT) (Masdeu et al., 2005), decreased metabolism in the parietotemporal association cortex, and precuneus nuoro-deoxyglucose (FDG) PET (Masdeu et al., 2005) and amyloid deposition on Pittsburgh compound B PET (Rowe et al. 2007). The frontotemporal dementias, or other lobar dementia syndromes, can be identified early on FDG PET by decreased metabolism in the affected region, which eventually also shows atrophy on MRI (Foster et al. 2007).

METHODS OF BRAIN MAPPING

Functional neuroimaging has assumed a central role in neuroscience research in the past two decades. Positron emission tomography (PET) and single photon emission computed tomography (SPECT) enable the visualization of blood flow or glucose metabolism in the resting state. Functional magnetic resonance imaging (fMRI) is used to measure brain activity in response to a stimulus or task. fMRI, in particular, allows insight into the physiology and pathology of perception and cognition and has become an indispensable tool for psychological, neurological, and psychiatric research. Its clinical applications, however, remain very limited at present. Its most important current use is for preoperative brain mapping. In psychiatry, fMRI has led to an improved understanding of the cerebral correlates of psychopathological phenomena (Linden & Deeks 2002; Kirche et al., 2004), cognitive disorders, and genetic risk factors. Studies on changes in brain activity during the treatment of mental illness may have particular clinical relevance. Functional neuroimaging has been used to study the effects of treatment, not just with medications of different kinds, but also with electroconvulsive therapy, vagus nerve stimulation (Zobel et al., 2005), as well as various types of psychotherapy. This use of psychological techniques (psychometry and psychotherapy) in combination with biological techniques (functional neuroimaging) typifies the trend in modern psychiatry toward the integration of these two approaches to mental illness, which were once widely held to be incompatible. Possible clinical applications of functional neuroimaging in psychiatry include assessment of the effects of treatment, determination of differential indications for treatment, identification of target areas for neurophysiological treatment methods, either invasive (deep brain stimulation) or noninvasive (transcranial magnetic stimulation), and development of neurobiologically inspired treatments such as neuropsychotherapy and neurofeedback.

Some specific techniques of functional brain imaging are discussed below

QEEG (QUANTITATIVE ELECTROENCEPHALOGRAM)

Quantitative electroencephalogram is a diagnostic tool, which measures electrical activity in the form of brainwave patterns. QEEG is defined as a method for analyzing the electrical activity of the brain to derive quantitative patterns that may correspond to diagnostic information and/ or cognitive deficits. It is a painless, non invasive procedure that takes around 30 minutes and helps in identification of neuropsychiatric problems. It enables the tracking of progress with different therapies - pharmacotherapy as well as psychotherapy. QEEG has been proposed to be useful in the diagnosis of various psychiatric disorders. However, the clinical significance of these distinctive patterns of brain wave activity is unknown. Both EEG and QEEG provide information about real time brain functioning at the time of evaluation. It is essential to understand that QEEG is different from EEG in that a standard EEG records electrical activity or brain waves that are representative of underlying cortical brain activity. A quantitative EEG (QEEG) on the other hand, applies mathematical and statistical analysis to brain waves and compares them in accordance with age and gender variables obtained through EEGs of people without any known brain dysfunction. The QEEG method helps in creating a map through this combination of precise measurement and quantitative measurement.

Evoked Potential: An evoked potential or evoked response is an electrical potential recorded from the nervous system after the presentation of a stimulus. Neuroelectrical responses to sensory stimuli can be readily and non invasively recorded using this averaging technique first employed by Dawson (1947). To generate EP record, the EEG may be used to measure brain activity following a sensory stimulus. The most widely used technique of this type involves the measurement of cortical response to a standard visual stimulus, a black and white chequer board that is moved in front of patient's eyes. Each movement of the chequerboard triggers a recording, and this process is repeated hundreds of times to enable an average response to be derives and recorded by computerised techniques. According to Walsh and Butler (2005), the clinical utility of EPs is based on their ability to:

- Demonstrate abnormal sensory system conduction, when the history and/ or neurological examination are equivocal
- Reveal subclinical involvement of a sensory system, i.e. a silent lesion
- Help define anatomic distribution
- Monitor changes in patient's neurological status

Although modality on theoretical basis one might assume that EPs may be conveniently conducted in all sensory modalities, however, in clinical practise, it is most commonly used with visual (Visual evoked potentials), Auditory modality (Auditory evoked potentials- AEP), brainstem auditory evoked responses (BSAER) and somatosensory evoked potentials (SSEPs).

POSITRON EMISSION TOMOGRAPHY (PET)

The functioning of brain depends on the level glucose provided by the circulating blood for 99% of its energy needs. For glucose to be used by the brain or any other organ, it must get from the blood to the cells that need it. Many studies show that the brain's demand for glucose depends on its work related requirements. It needs more when working hard, as is the case when trying to remember something. Many brain-imaging techniques are available today. However, only the Positron Emission Tomography (PET) and SPECT scan can detect areas of decreased or increased brain glucose metabolism.

PET and SPECT are most commonly used techniques to measure regional blood flow or regional cerebral glucose metabolism. PET and SPECT enable the study of specific chemicals such as receptors and transporters that are in very low concentrations in the brain. Receptors are crucial in brain functioning as the primary effect or sites of neurotransmission on postsynaptic membranes for endogeneous neurotransmitters and exogeneously administered drugs. PET and SPECT provide the revolutionary opportunity to image and study these specific chemical brain sites.

PET scan is an imaging technique that uses a very small dose of radioactive substance/s called radiotracer, to help image and study the functioning of various organs and tissues. A tracer is essentially a biological compound of interest labelled with a positron emitting isotope, such as 11C, 18F, and 15O. The radioactive tracers used for PET are specially synthesized molecules which contain positron emitting isotope of common elements like carbon, oxygen, nitrogen and fluoride. These isotopes are used because they have relatively short half-lives (minutes to less than two hours), allowing the tracers to reach equilibrium in the body, but without exposing the subjects to prolonged periods of radiation. Further, isotopes of these elements are chosen over isotopes of other elements because these elements also naturally occur in many biological molecules (Kapur, 1994). The tracer is administered either as an injectable or inhalant. The amount of tracer administered, the time of administration and duration of uptake of tracer all vary depending upon the nature of tracer and the purpose of study. For carrying out PET scan, the cyclotron accelerates a beam of protons using two high voltage electrodes and directs it towards the target nuclei, thereby incorporating an extra proton into them. This generates new isotopes with a neutron-to-proton ratio which are energetically unstable. The isotopes are then combined with the tracer. Most of these unstable isotopes undergo a process of decay whereby the excess proton is converted into a neutron, a positron, and a neutrino. The emitted positron and electron travel in opposite directions up to a range of a few millimetres in tissue before being annihilated mutually. This mutual annihilation process produces two photons of equal energy that travel in opposite directions.PET scanners contain several rings of scintillation detectors that are usually made of bismuth germanate (BGO) (Tai & Piccini, 2004). The pair of photons (travelling in opposite directions) produced from a single annihilation, register almost simultaneously on opposing pairs of BGO detectors as a "coincidence event". The paths of these two photons, are termed as lines of response (LORs), and are thus traced and recorded. The rings of BGO detectors register thousands of coincidence events emitted from the subject per second. The data gathered from the coincidence events and LORs are used to determine the source of positron annihilation at any given time. These are then converted into a tomographic image using standard reconstruction software (Cherry, 1996). Co-registration of PET images to high resolution magnetic resonance images (MRI) enables the accurate anatomical localisation of functional changes displayed on PET (Tai et al., 2004).

Once administered to the patient, the tracer is distributed to various body parts. In order to study the kinetics of the tracer accurately, it is utmost important to know its plasma concentration (often by continuous arterial blood sampling to obtain arterial input function) and cerebral concentration (from dynamic sequences of PET images) over time, correcting for metabolism and radioactive decay of the compound (Tai et al., 2004).

Specific tracers are chosen to illustrate the particular brain functions the investigators are interested in. For example, 18F-2-deoxyglucose (18FDG) is used to investigate cerebral glucose metabolism, whereas H215O is used to examine cerebral blood flow. PET scans may be carried out at rest, while performing certain tasks or after administration of challenge compounds. Several recent studies have suggested that using the principle of competition between endogenous neurotransmitters and administered tracers for the same neuroreceptors, it is possible to image acute fluctuations in the concentration of synaptic neurotransmitters (Laruelle, 2000).

The societies for Nuclear Medicine (Waxman et al., 2009) have issued guidelines for patient preparation for PET scan. Some of these are listed below in text box:-

Patient Preparation

Pre-arrival (Patient Instruction)

1. Patient should be fasting for 4 - 6 hours.
2. Oral hydration with water is encouraged.
3. Avoid caffeine, alcohol, or drugs that may affect cerebral glucose metabolism.
4. Required medications should be taken with water.
5. Intravenous fluid containing dextrose or parenteral feeding should be withheld for 4-6 hours.

Pre-Injection

Preparation:

Environment – Should be stable for at least 30 minutes prior to FDG injection and in the subsequent uptake phase (at least 30 minutes).

1. The patient should be placed in a quiet, dimly-lit room.
2. Background noise is acceptable with eyes open.
3. Alternatively, studies can be performed with eyes closed if this is being done consistently across comparable scans.
4. The patient should be seated or reclining comfortably.
5. Place intravenous access at least 10 minutes prior to injection to permit accommodation.
6. Instruct the patient to relax, not to speak or read and to avoid major movements.
7. Minimize interaction with the patient prior to, during, and up to at least 30 minutes post injection.

Blood Glucose: Should be checked prior to FDG administration.

1. If greater than 150 - 200 mg/dL, most institutions reschedule the patient. When hyperglycemia is present, there is increased competition of elevated plasma glucose with FDG at the carrier enzyme and is often usually associated with high intracellular glucose and circulating insulin levels driving FDG into muscle resulting in reduced uptake in the brain.
2. Lowering serum glucose by administration of insulin can be considered, but the administration of FDG should be delayed following insulin administration (with the duration of the delay dependent on the type and route of administration of insulin), probably because the correction of increased intracellular glucose levels lags behind the correction of the plasma glucose level.

Bladder

1. Before the scanning procedure is started, the patient should void the bladder for maximum comfort during the study.
2. Advise the patient to void again after the scanning session to minimize radiation exposure.

EEG

2. For preoperative evaluation of epilepsy, continuous EEG recording is recommended.
3. Monitoring should start ideally 2 hours before injection in order to ensure that FDG is not administered in a post-ictal state and should be maintained at least until 20 minutes post-injection. For adequate image interpretation, it is important to be aware if seizures occurred prior to imaging.

**Patient preparation guidelines for PET scan taken from guidelines (2009) as given by society of nuclear medicine procedure guideline for FDG PET brain imaging

SINGLE PHOTON EMISSION COMPUTED TOMOGRAPHY (SPECT)

Single Photon Emission Computed Tomographyis a functional nuclear imaging technique that allows visualization and quantification of different invivo physiological and pathological features of brain neurobiology (Joseph, 2007). The SPECT was introduced in the early 1980s with the purpose of evaluation of regional cererbral perfusion and receptor density studies (Masdeu, 2007). Brain SPECT can be used to define a patient's pathological status when neurological or psychiatric symptoms cannot be explained by structural neuroimaging findings (Santra et al., 2014). Though there is ample evidence in the literature validating brain SPECT as a promising technique to track human behavior and correlating psychiatric disorders with dysfunction of specific brain regions, it is rarely utilized technique in routine psychiatric practice (Santra, et al., 2014). In SEPCT, a flow tracer or a receptor binding substance is tagged with a radio nuclide and injected intravenously. It is based on the assumption that the flow tracer accumulates in different areas of the brain proportionally to the rate of delivery of nutrients to that volume of brain tissue (Masdeu, 2007). SPECT images are generated using gamma cameras or ring type imaging systems that record photons emitted by the tracer trapped in the brain. SPECT results in better image quality than the traditional two dimensional or planar imaging techniques because focal sources of activity are not superimposed on one another. As a result of this, the contrast between the target and the background is greatly increased. In general, the system cost for SPECT depends directly on the number of camera heads and crystals. Scanning time in SPECT depends on the imaging system, the type of radiopharmaceutical element used and the type and quality of image desired. High resolution images of the whole brain can be obtained in approximately 20 to 30 minutes. The volume imaging capacity of most SPECT systems permits reconstruction at any angel- including the axial, coronal, and saggital planes, or even at same angels as obtained by CT or MRI techniques which also enables and helps in effective comparison of findings from various imaging techniques. Like most imaging techniques, SPECT also require some patient preparation. As outlined by Santra and Kumar (2014), before arrival, patients should be instructed to avoid, if possible, caffcinc, alcohol, or other drugs known to affect cerebral blood flow (CBF). Since, brain perfusion is sensitive to neuronal activities, hence, tracer injection should be done in a quiet room and no interaction with patients (neither with the hospital staff nor caregivers) at this time is desirable, so as to avoid any sensorial and cognitive stimulation. To avoid head movement during scanning (20-30 min), the patient should be comfortable and relaxed. The uncooperative patients (those with severe cognitive impairment or with loss of insight) may need sedation (Santara, 2014). Tracer injection must precede sedation to avoid sedation-induced blood flow changes. Appropriate positioning is needed to keep the collimators as close as possible to the patient's head and to get entire brain within the center of field of view (Juni et al., 2009; Catafau, 2001).

FUNCTIONAL MAGNETIC RESONANCE IMAGING (f-MRI)

Functional Magnetic Resonance Imaging measures the small changes in blood flow that occur with brain activity. fMRI is usually able to detect abnormalities within the brain that cannot be detected with other imaging techniques. In MRI, patient is placed in strong magnetic field. Under the influence of magnetic field, protons that have magnetic field of their own and hence a polarity line up and orient in accordance with the forces in the field. Conventional MRI is used largely for radiological diagnosis and produces spatial maps of the properties of single photons that are contained in water molecules. Conventional MRI reveal anatomic details with high resolution in three dimensions and are able to differentiate soft tissues very well. The contrast that is revealed within images is largely due to variations in the density of water within the tissues and in the manner in which water molecules interact with macromolecules. Functional MRI uses similar imaging techniques and same equipment as conventional MRI. f-MRI relies on detecting small changes in the signals used to produce magnetic resonance images that are associated with neuronal activity in the brain. f-MRI is non invasive, safe and can be repeated without any known risks in both children as well as adults. f-MRI detects the blood oxygen level dependent changes in the MRI signal that arise when changes in neuronal activity occur following a change in brain state which are produced by a stimulus or a task. It is based on the principle that various behaviours and brain functions rely on coordinated interaction of various components of the brain.

CONCLUSION

Neuroimaging plays a crucial role in psychiatry and mental health as it can potentially be used to identify biomarkers of a disorder, prognosis or treatment, elucidate biological pathways along with informing and monitoring newer therapies. Early and prophylactic interventions present an emerging future direction in clinical psychiatry (McGory et al., 2011) and neuroimaging has the potential to help in identification of individuals at risk and monitor the effects of intervention.

REFERENCES

Agarwal, N., Port, J. D., Bazzocchi, M., & Renshaw, P. F. (2010). Update on the use of MR for assessment and diagnosis of psychiatric diseases. *Radiology*, *255*(1), 23–41. doi:10.1148/radiol.09090339 PMID:20308442

Albert, D. J., Walsh, M. L., & Jonik, R. H. (1993). Aggression in humans: What is its biological foundation? *Neuroscience and Biobehavioral Reviews*, *17*(4), 405–425. doi:10.1016/S0149-7634(05)80117-4 PMID:8309650

Boone, K. B., Miller, B. L., Rosenberg, L., Durazo, A., McIntyre, H., & Weil, M. (1988). Neuropsychological and behavioral abnormalities in an adolescent with frontal lobe seizures. *Neurology*, *38*(4), 583–586. doi:10.1212/WNL.38.4.583 PMID:3352916

Bunevicius, A., Deltuva, V. P., Deltuviene, D., Tamasauskas, A., & Bunevicius, R. (2008). Brain lesions manifesting as psychiatric disorders: Eight cases. *CNS Spectrums*, *13*(11), 950–958. doi:10.1017/S1092852900014000 PMID:19037181

Catafau, A. M. (2001). Brain SPECT in clinical practice. Part I: Perfusion. *Journal of Nuclear Medicine*, *42*, 259–271. PMID:11216525

Cherry, S. R., & Phelps, M. E. (1996). Imaging brain function with positron emission tomography. In A. W. Toga & J. C. Mazziotta (Eds.), *Brain Mapping—The Methods* (pp. 191–222). San Diego, CA: Academic Press.

Dawson, G. D. (1947). Investigations on a patient subject to myoclonic seizures after sensory stimulation. *Journal of Neurology, Neurosurgery, and Psychiatry, 10*(4), 141–162. doi:10.1136/jnnp.10.4.141 PMID:18905644

Gelder, M. J. (1996). Biological Psychiatry in perspective. *British Medical Bulletin, 52*(3), 401–407. doi:10.1093/oxfordjournals.bmb.a011556 PMID:8949246

Grafman, J., Vance, S. C., Weingartner, H., Salazar, A. M., & Amin, D. (1986). The effects of lateralized frontal lesions on mood regulation. *Brain, 109*(6), 1127–1148. doi:10.1093/brain/109.6.1127 PMID:3790971

Hariri, A. R., & Weinberger, D. R. (2003). Imaging Genomics. *British Medical Bulletin, 65*(1), 259–270. doi:10.1093/bmb/65.1.259 PMID:12697630

Juni, J. E., Waxman, A. D., Devous, M. D., Tikofsy, R. S., Ichise, M., VanHeertum, R. L., & … . (2009). Procedure guideline for brain perfusion SPECT Using 99mTc Radiopharmaceuticals 3.0. *Journal of Nuclear Medicine Technology, 37*(3), 191–195. doi:10.2967/jnmt.109.067850 PMID:19692453

Kapur, S., Houle, S., & Brown, M. G. (1994). Positron Emission Tomographhy in Psychiatry: New Sights, New Insights. *Jefferson Journal of Psychiatry, 12*(1), 58–74. doi:10.29046/JJP.012.1.007

Kircher, T., Schneider, F., Sauer, H., & Buchkremer, G. (2004). Funktionelle Bildgebung am Beispiel der Schizophrenie. *Deutsches Arzteblatt, 101*, A–1975.

Laruelle, M. (2000). Imaging synaptic neurotransmission with in vivo binding competition techniques: A critical review. *Journal of Cerebral Blood Flow and Metabolism, 20*(3), 423–451. doi:10.1097/00004647-200003000-00001 PMID:10724107

Linden, D. J. (2012). The challenges and promises of neuroimaging in Psychiatry. *Neuron, 73*(1), 8–22. doi:10.1016/j.neuron.2011.12.014 PMID:22243743

Linden, D. J., & Dierks, T. (2002). Funktionelle Bildgebung akustischer Halluzinationen bei der Schizophrenie. *Nervenheilkunde: Zeitschrift für Interdisziplinaere Fortbildung, 21*, 331–335.

Masdeu, J. C. (2011). Neuroimaging in Psychiatric Disorders. Neurotherapeutics. *The Journal of the American Society for Experimantal Neurotherapeutics., 8*(1), 93–102. doi:10.100713311-010-0006-0 PMID:21274689

Masdeu, J. C., Zubieta, J. L., & Arbizu, J. (2005). Neuroimaging as a marker of the onset and progression of Alzheimer's disease. *Journal of the Neurological Sciences*, *236*(1-2), 55–64. doi:10.1016/j.jns.2005.05.001 PMID:15961110

Moise, D., & Madhusoodanan, S. (2006). Psychiatric symptoms associated with brain tumors: A clinical enigma. *CNS Spectrums*, *11*(1), 28–31. doi:10.1017/S1092852900024135 PMID:16400253

Rowe, C. C., & Ackermann, U. (2007). Imaging beta-amyloid burden in aging and dementia. *Neurology*, *68*(20), 1718–172. doi:10.1212/01.wnl.0000261919.22630.ea PMID:17502554

Santra, A., & Kumar, R. (2014). Brain perfusion single photon emission computed tomography in major psychiatric disorders: From basics to clinical practice. *Indian Journal of Nuclear Medicine*, *29*(4), 210–221. doi:10.4103/0972-3919.142622 PMID:25400359

Sherwin, I., Peron-Magnan, P., Bancaud, J., Bonis, A., & Talairach, J. (1982). Prevalence of psychosis in epilepsy as a function of the laterality of the epileptogenic lesion. *Archives of Neurology*, *39*(10), 621–625. doi:10.1001/archneur.1982.00510220019004 PMID:6812552

Tai, Y. F., & Piccini, P. (2004). Applications of positron emission tomography (PET) in neurology. *Journal of Neurology, Neurosurgery, and Psychiatry*, *75*(5), 669–676. doi:10.1136/jnnp.2003.028175 PMID:15090557

Walsh, P., Kane, N., & Butler, S. (2005). The clinical role of evoked potentials. *Journal of Neurology, Neurosurgery, and Psychiatry*, *76*(suppl_2), 16–22. doi:10.1136/jnnp.2005.068130 PMID:15961863

Waxman, A.D., Karl, H., David, H., Peter, H., Satoshi, M., & Masanori, I. (2009). *Procedure Guideline for FDG-PET Brain Imaging v1.01*. Society of Nuclear Medicine Procedure Guideline for FDG PET Brain Imaging Version 1.0.

Zobel, A., Joe, A., Freymann, N., Clusmann, H., Schramm, J., Reinhardt, M., ... Broich, K. (2005). Changes in regional cerebral blood flow by therapeutic vagus nerve stimulation in depression: An exploratory approach. *Psychiatry Research*, *139*(3), 165–179. doi:10.1016/j.pscychresns.2005.02.010 PMID:16043331

KEY TERMS AND DEFINITIONS

EP: An evoked potential (EP)or evoked response is an electrical potential recorded from the nervous system after the presentation of a stimulus. Neuroelectrical responses to sensory stimuli can be readily and non-invasively recorded using this averaging technique first employed by Dawson (1947). To generate EP record, the EEG may be used to measure brain activity following a sensory stimulus.

f-MRI: Functional magnetic resonance imaging (f-MRI) measures the small changes in blood flow that occur with brain activity. fMRI is usually able to detect abnormalities within the brain that cannot be detected with other imaging techniques.

Neuroimaging: Neuroimaging or brain imaging is the use of various techniques to either directly or indirectly image the structure, function, or pharmacology of the nervous system. It is a relatively new discipline within medicine, neuroscience and psychology. Neuroimaging is the discipline that deals with the *in vivo* depiction of anatomy and function of the central nervous system (CNS) in health and disease.

PET: PET scan is an imaging technique that uses a very small dose of radioactive substance/s called radiotracer, to help image and study the functioning of various organs and tissues. The radioactive tracers used for PET are specially synthesized molecules which contain positron emitting isotope of common elements like carbon, oxygen, nitrogen and fluoride.

QEEG: Quantitative electroencephalogram (QEEG) is a diagnostic tool, which measures electrical activity in the form of brainwave patterns. QEEG is defined as a method for analyzing the electrical activity of the brain to derive quantitative patterns that may correspond to diagnostic information and/ or cognitive deficits. It is a painless, noninvasive procedure that takes around 30 minutes and helps in identification of neuropsychiatric problems.

SPECT: Single photon emission computed tomography is a functional nuclear imaging technique that allows visualization and quantification of different in vivo physiological and pathological features of brain neurobiology. Brain SPECT can be used to define a patient's pathological status when neurological or psychiatric symptoms cannot be explained by structural neuroimaging findings.

Chapter 12
Neuropsychological Assessment

Manoj Kumar Bajaj
https://orcid.org/0000-0002-9999-7888
Government Medical College and Hospital, Chandigar, India

ABSTRACT

This chapter will cover the basic concepts of neuropsychological assessment. This chapter will cover the main issues involved while assessing the person with brain damage. The process of neuropsychological assessment, relevant history for the purpose of comprehensive reporting of findings and responding to the referral questions, better understating of the goals of assessment, selection of the assessment approaches, issues involved in administration of the tests, theories involved while interpreting the test findings, and reporting of findings of the assessment will help the reader to understand holistic view of the individual's capacities.

CHAPTER OUTLINE

- Introduction
- History of neuropsychological assessment
- Types of neuropsychological assessment using in India
- Goals of Neuropsychological Assessment
- Approaches Neuropsychological assessment
- Neuropsychological Test Findings Interpretation
- Neuropsychological assessment process
- Conclusion

DOI: 10.4018/978-1-7998-2860-0.ch012

INTRODUCTION

Neuropsychological assessment refers to the measurement of cognitive functions and processes with the aim to establish whether cognitive dysfunction or cognitive impairment is present in individuals, typically in patients with (suspected) brain disease, psychiatric disorder or information-processing complaints. Neuropsychological assessment is preferably performed by clinical neuropsychologists who have expertise in the study of the neural correlates of behavior and cognition (Kessels, 2016).

The neuropsychological evaluation involves an interview and the administration of tests. Neuropsychological tests (unlike bedside cognitive and behavioural neurologic screens) are standardized, meaning that they are given in the same manner to all patients and scored in a similar manner time after time. An individual's scores on tests are interpreted by comparing their score to that of healthy individuals of a similar demographic background (i.e., of similar age, education, gender, and/or ethnic background) and to expected levels of functioning. In this way, a neuropsychologist can determine whether one's performance on any given task represents a strength or weakness. Although individual scores are important, the neuropsychologist looks at all of the data from the evaluation to determine a pattern of cognitive strengths and weaknesses and, in turn, to understand more about how the brain is functioning.

Neuropsychological tests evaluate functioning in a number of areas including: intelligence, executive functions (such as thinking, planning, abstraction, set shifting, response inhibition, problem solving, and working memory), attention, memory, language, perception, sensory motor functions, motivation, mood state and emotion, quality of life, and personality changes. The areas addressed in an individual's evaluation are determined by the referral question (what the referring doctor and patient wants to know), patient's complaints and symptoms, and observations made during interview and test administration.

The tests are typically pencil and paper type tests. Some tasks might be self-reports meaning that they are completed by the patient with assistance from a technician, but the majority of the tests require administration by a neuropsychologist or trained, skilled clinical psychologists.

SHORT HISTORY OF NEUROPSYCHOLOGICAL ASSESSMENT

Neuropsychological processes are very much evident from a very early period in history, the first attempt to localize cognitive facilities in the ventricles is generally attributed to Herophilius around 300 B.C. (Mann, 1979). Huarte (1529-1588) first proposed a localizationist position involving both ventricles and brain tissue (Hunt, 1869), and a century later Willis (1620-1675) made the complete translation of localization of function to the cerebrum (McHenry, 1969). Localization attempts during this period almost universally focused on finding a single area of the brain underlying all mental functions (Luria, 1966). Localization of specific functions by specific cortical areas was recognized early in the eighteenth century, and de Petit (1644-1771) demonstrated contralateral innervation in both man and dogs.

In the 17th and 18th century literature dealing especially with language disorders following a brain disorder, one finds observations of physicians of striking dissociations of mental faculties that were impaired while others remained intact. Although aphasia had been described more than a century earlier by Genner, a French physician named Paul Broca (1824-1880) was the first (1861) to report detailed autopsy findings demonstrating the localization of speech in the third frontal convolution later which called Broca's area. By 1863 Broca and his colleagues had collected 19 cases of aphasia which all showed lesions in the third frontal convolution of the left cerebral hemispheres. Although Broca's impressive findings did not fully convert the critics of localization of function, they were influential in shaping the course of research of a number of important scientists of the day. Around the middle of the 19th century, neuropsychiatrists like Carl Wernicke began to develop procedures for assessing more specific components of mental functioning. German physicians, Conrad Rieger and Theodor Ziehen, seem to have developed the first neuropsychological test batteries. Kurt Goldstein, inspired by the rising Gestalt theory, argued that not the test score but the strategy used by a patient to perform a task is important(Golden, 1983).

Alexander Luria also promoted an approach to assessment that was mainly based on subjective judgment. Luria's innovative ideas that the brain underlies the ability to carry out goal-directed behaviors and is shaped by environmental and cultural contexts (e.g., language) led to his systematic characterization of functional brain systems (Luria, 1966). Luria stimulated appreciation of brain specialization and moved away from the one-size-fits-all diagnosis of brain disease; however, his techniques were highly flexible and non standardized, making it difficult to reliably reproduce across examiners and patients. Ultimately, Charles Golden further standardized and combined the works of both Luria and Christensen in his development of the Luria-Nebraska Neuropsychological Battery (LNNB; Golden, Purisch, & Hammeke, 1979), scaling down almost 2000 original measurement items to 269 items covering 14 scales (e.g., motor, rhythm, memory, intelligence; (Purisch, 2001).

Studies on individual differences led to the development of an intelligence test battery by Alfred Binet. This battery was later transformed into the Army Alpha and Army Beta tests for selecting soldiers. Components of these intelligence tests have survived in the test kit of the modern neuropsychologist. This tradition also stimulated the development of psychometric analysis of tests. Two pioneers in the field of neuropsychological assessment were Shepherd Ivory Franz, favoring a clinical approach, and Ward Halstead, stimulating a strongly psychometric-based approach. The evaluation of language disorders has always been a specific area, requiring its own set of tests. The first comprehensive language battery was compiled by Bastian. Around the middle of the 20th century, when the localization of function approach had been rejected, neurologists preferred to examine language disorders clinically, using a battery that evaluated speech, comprehension, reading, and writing.

TYPES OF NEUROPSYCHOLOGICAL ASSESSMENT

List of standardized neuropsychological tests were presented in foundation of neuropsychology (see chapter No.7)

In India: Neuropsychological tests were developed and currently using in India: There are number of psychological tests batteries have been developed/ standardized in India using concept of dysfunction. A few of them are based on well recognized concept of higher brain functions like functioning intelligence, estimation of past intelligence, memory functioning, perceptual acuity etc (Pershad & Verma, 1990). Others are based on neurological examinations and follows evaluation of functioning of hemisphere and lobes (Gupta, 2000 & Rao et al, 2004). Some of the significant test batteries published during the last about two decades in India are as follows (Vaishali, 2019).

1. **PGI Battery of Brain Dysfunction:** By Dwarka Pershad and Verma (1990 & 2007). This battery includes five tests viz. Bhatia's Short Scale, Verbal Adult Intelligence Scale, PGI Memory Scale, Bender Visual Motor Gestalt Test and, Nahor Benson Test of Perceptual Acuity. All these tests are fully validated and widely acclaimed throughout the world. Norms are developed for 20 to 50 years age group; following factorial sampling design of 2 X 3 X 5 [two levels of sex, three levels of education and five levels of age]. It gives global rating of cognitive dysfunction based on 19 test variables and estimates well accepted/validated psychological concepts of (a) intelligence, (b) memory and (c) gestalt formation or perceptual acuity. It provides a profile of current cognitive functioning of the subject.
2. **AIIMS Comprehensive Neuropsychological Battery in Hindi [Adult Form]:** by Surya Gupta et al. (2000). This comprehensive tests battery is largely based on Luria-Nebraska Neuropsychological Battery and potentially useful for both diagnosis and rehabilitation. It consists of thirteen basic scales- Motor Scale, Tactile Scale, Visual Scale, Receptive

Speech Scale, Expressive Speech Scale, Reading Scale, Writing Scale, Arithmetic Scale, Memory Scale, Intellectual Processes Scale, Left Hemisphere Scale, Right Hemisphere Scale, Pathognomonic Scale and Total Score Scale. It consists of 160 items. It is claimed that it is an ideal psychological instrument for clinical and research settings facilitating identification, lateralization and localization of brain lesions, in the subjects from 15 to 80 years of age.

3. **NIMHANS Neuropsychology Battery - 2004:** by Rao et al. (2004). There are 19 tests in all, which measures 15 functions, in seven neuropsychological domains such as: Speed, Attention, Executive functions, Comprehension, Verbal learning and memory, Visuo-spatial construction and, Visual learning & memory. The sub tests used in the battery include,[a] Finger Tapping Test [b] Digit Symbol Substitution [c] Color Trails [d] Digit Vigilance [e] Traids Test [f] Controlled Oral Word Test [g] Association Test [h] Animal Names Test [i] Design Fluency Test [j] N Back Test [verbal and visual] [k] Self Ordered Pointing Test [l] Tower of London Test [m] Wisconsin Card Sorting Test [n] Stroop Test [o] Token Test [p[Auditory Verbal Learning Test [q] Passages Test [r] Complex Figure Test and [s] Design Learning Test. Normative data has been established for adults in the age range of 16 to 65 years following a factorial sampling design. A separate manual has been published under the name of NIMHANS Neuropsychological Battery for Children by Kar, et al (2004) for the children in the age range of 5-15 years.

GOALS OF NEUROPSYCHOLOGICAL ASSESSMENT

With the advent of increasingly sensitive and multimodal neurologic biomarker data, neuropsychological assessment shifted from its original role of localization and lateralization to functional aspect of the brain functioning. For the planning of rehabilitation in traumatic brain injuries, to overcome impact of various psychotropic drugs causing cognitive disabilities, assessment of disability for the benefits offered from various schemes and insurance claims. Another important function has been confirmed by the detailed neuropsychological assessments for candidates to decide for the eligibility for epileptic surgeries in refractory epilepsy diagnosis. Goals for the neuropsychological assessment are based on specific referral questions and needs according to affected individuals and their families.

- Diagnostic information for detection of dementias or other traumatic conditions (e.g., to determine whether possible mental and behavioural changes are related to the movement disorder, depression, another brain disease or treatment)
- Measurement of recovery of functioning to provide a baseline against which subsequent evaluations can be compared. Thereby your clinician can decide whether your functioning has declined because of the disease process or document whether your functioning has worsened or improved as a result of treatment
- Measurement of treatment response (evaluation before and after functional neurosurgical procedures (e.g., deep brain stimulation) or pharmacological treatment to help determine if a given treatment is appropriate for a particular person and whether treatment has had any positive or negative effects on mental functions and behaviour.
- Reveal areas of daily functioning (e.g., financial management) with which the patient may need assistance indicate rehabilitation potential. For example, will the individual benefit from certain cognitive or behavioural treatments, occupational therapy, or pharmacotherapy.

APPROACHES OF NEUROPSYCHOLOGICAL ASSESSMENT

Various assessment approaches has been advocated as per the purpose of the referral questions by the various medical experts e.g. the case may be referred for the neuropsychological rehabilitation purpose after head injury, then depending upon the site and extent of the lesion approaches to assessment is decided by the neuropsychologist.

Flexible Battery: One assessment style that naturally extended from clinicians' early instincts to describe observed behaviours was the flexible integration of standardized measures. In the pure flexible battery approach, neuropsychologists administer only those measures directly related to the patient's presenting symptoms (e.g., multiple memory measures to characterize memory symptoms). Consequently, the administered battery inherently changes with a patient's referral question, while the examiner pursues the aim of being both as efficient and sensitive as possible to the presenting problem (Benton, 1994; Lezak, 1976).

Fixed Battery Approach: An entirely distinct assessment approach is used when neuromedical history, or clinically apparent syndrome is to be established, in this relatively comprehensive set of measures administered to all patients regardless of presenting symptoms is been used. Although lengthier and requiring greater burden and expense, fixed batteries maintain a constant testing condition, allowing for direct comparison of measures across disparate neurological disorders. Given their highly standard nature, research regarding fixed batteries is more easily facilitated, has garnered more comprehensive normative data, better understanding of their psychometric properties, and a broader empirical base from which to draw interpretations.

Two major leaders in the history of neuropsychological assessment pioneered the fixed battery approach, Ward Halstead and Ralph Reitan (Reitan & Davidson, 1974). Halstead was a physiological psychologist in the mid-twentieth century and believed that there was no science in the individual event, and emphasized the need to develop exact, systematic procedures with sufficient comparison cases to interpret an individual test score (Reitan, 1994).

Boston Process Approach: Another important turning point in neuropsychological practices was the development of what is now referred to as the "Boston Process Approach" led by Edith Kaplan (Kaplan, 1988). Although the Boston approach can be used with either flexible or fixed batteries, it is most commonly associated with and perhaps conceptually suitable to more flexible evaluations. The Boston approach places emphasis on how the patient comes to an answer (e.g., types of errors committed) rather than reliance on a single objective score. In this approach, testing the limits of a patient's cognitive abilities to elicit behaviours that may not traditionally present during standardized testing is emphasized.

Flexible Evaluation Approach: To avoid misconceptions, the pure flexible battery differs from the *flexible evaluation* approach that is used by over three-quarters of contemporary clinical neuropsychologists (Larrabee, 2008). Current common practice represents aspects of both flexible and fixed approaches, such that a fairly standardized, fixed set of measures is given to most patients with some flexibility to add or subtract measures given the specific referral question (Bigler, 2007).

NEUROPSYCHOLOGICAL TEST FINDINGS INTERPRETATION

Neuropsychological assessment serves various purposes one of them is the planning rehabilitation post injury to brain. After administration and scoring of a set of neuropsychological tests, and the conversion of raw scores into standard scores, the results have to be interpreted from a clinical perspective. At the level of a single test, classification can be done using the standard scores, but note that the cut-off scores are based on convention rather than external criteria. There are various limitations in the interpretation of the scores and comparison to their respective age and education group. Every individual has its own spectrum of functioning on various domains.

Clinical interpretation also includes interpretation of other data that are available to a neuropsychologist, such as an individual's medical history, findings of magnetic resonance imaging, qualitative observations of test behavior made during the assessment and the person's intellectual and cognitive functioning before the onset of the cognitive complaints, brain disease or disorder.

It is to be noted that diagnosis is rarely made on the basis of a single test performance; it is the result of the aggregate interpretation of multiple neuropsychological tests, test behavior, information from the clinical interview with the patient and medical history. Author advocates the sound knowledge and understanding of the processing of the brain functions, as brain simultaneously process information in various parts of the brain using various connections and pathways and results into a behaviour. So, it is advisable while interpreting the findings of the neuropsychological tests only test scores are not the actual dysfunctions in that particular individual. Collaborating with their daily life difficulties will help us understanding that merely scores are not truly representative of their difficulties or dysfunctions in the daily life.

Also, apparent inconsistencies findings may be present in patients profile (e.g., adequate performance on verbal memory test but an impaired performance on working memory test), which can make the interpretation of the test results difficult. It is important to keep in mind that a poor test performance is not a direct proof of brain dysfunction; as such a performance can also be affected by other factors, such as developmental disorders, personality changes, depressed mood, inability to comprehend complex test instructions, lack of interest and motivation. Lack of motivation may result in insufficient mental effort being displayed during neuropsychological assessment, affecting the symptom validity of the assessment (i.e., poor test performances that do not reflect the actual level of cognitive abilities of that individual). Specific tests exist to measure symptom validity, which can detect underperformance because of lack of motivation or because of feigned cognitive symptoms (malingering, which can be seen in some compensation-seeking individuals). Note also that subjective reports of cognitive complaints are not necessarily substantiated by an impaired performance on neuropsychological tests, but that such complaints can also be the result of rumination or worrying, false beliefs that one's memory has to be perfect all the time or psychological distress. In clinical practice, interpretation of neuropsychological test results requires knowledge of cognitive theories on brain– behavior function, understanding of normal and pathological development, knowledge of neurological diseases and mental disorders, experience with the neuropsychological tests at hand and their psychometric properties as well as understanding of the quality of the available normative data. Neuropsychologists usually describe the interpretation of the results based on certain theories of brain functioning such as "Double Dissociation" and their main conclusion in a neuropsychological report, which may support the medical diagnostic process, or guide treatment, care or other interventions.

"Double Dissociation" a term that was introduced by Hans-Lukas Teuber in 1955, refers to documenting two distinct patterns of impairment in two different groups or individuals, proving that two functions are neurologically distinct. In cognitive neuroscience, double dissociation is an experimental technique by which two areas of neocortex are functionally dissociated by two behavioural tests, each test being affected by a lesion in one zone and not the other. In a series of patients with traumatic brain injury, one might find two patients, A and B. Patient A has difficulty performing cognitive tests for, say auditory memory but has no problem with visual memory. Patient B has the opposite problem. By using neuroimaging (or neuropathology post-mortem) to identify the overlap and dissociation between lesioned areas of the brain, one can infer something about the localization of visual and auditory function in the normal brain.

NEUROPSYCHOLOGICAL ASSESSMENT PROCESS

Obtaining the important medical history, sequence of insult to brain, onset of the problem whether it is started abruptly, acutely or has insidious onset is the *first* step in the process of neuropsychological assessment. Age, educational status, occupational skills, economic status, handedness are very important to be recorded. It is very important to document the status of his brain functioning on Glasgow Coma Scale (GCS) at the time of insult to the brain and at the time of intake if patient presents with the confusion. Extent and location of the injury, disorder specific pointers, progress or deterioration, behavioural change, emotional change, changes in activity of daily living instrumental activities of daily living, disability or handicap along with the development of the problem premorbid functioning background data and change in behaviour and other specific investigations and findings on imaging such as MRI, CT, PET scan etc are very relevant and needs to be collected. It is worth pertinent to mention to record the neurological signs and any use of drugs and treatment conservative or operative to be noted. After collecting the relevant information mentioned above proceeding for the *second* step for planning the patient for Neuropsychological assessment. On the basis of the available observations and clinical record, it can be decided by the neuropsychologists that which assessment approach i.e. fixed or flexible battery approach or flexible evaluation approach is right for him depending upon the referral questions and otherwise. Here selection of the tests and its relevance to the referral question is very important to be emphasised along with its psychometric properties according to the age and education. It can be two tire approach in which patient is first evaluated for the basic functions of attention, orientation, memory and language comprehension. Later on if these are available then higher function such as executive functions, abstraction, problem solving can be assessed. Before assessing the basic functioning it is very important to ensure the various sensory functions such as vision or hearing is intact or accordingly plan assessment. The *third* step is to administration of the tests, in which important to note that patient's behaviour while assessment to be recorded on all subtests separately and clearly. Also record the time exceeding in completion of the test, implicit or explicit methods for responding to specific tests, trial and error, inability to extent the limbs, or motor weakness, attention level, arousal, motivation, comprehension of commands or verbal information, ability to hold information to respond, speech output, responses with cues etc. This particular information is very important when analysing the performance of each subtest to reach on a conclusion. *Forth* step in the neuropsychological assessment is interpretation of findings and scores using the double dissociation and componential analysis approach to reach on a conclusion. Synthesizing the obtain findings and summarising then according to their referral question. *Fifth* and the final step is to reporting the findings, after analysing and synthesising of the all information according to the steps described and addressing to the referral questions. Findings are being presented with the appropriate recommendations and likelihood of improvement in the respective functions. It is also important to note, analysed the findings relative to the pre morbid functioning obtained through history or any specific tests of estimating the premorbid functioning.

CONCLUSION

Neuropsychological assessment has multiple applications ranging from diagnosing to planning rehabilitation. A careful evaluation is warranted by a Neuropsychologist using in depth understanding of interpreting the obtained findings on various tests. With the innovation in the imaging science the role of neuropsychological assessment is changing from localization/ lateralization to assessing the future outcomes of the insult to brain, its recovery, functionality predictions and rehabilitation planning.

REFERENCES

Benton, A. L. (1994). Neuropsychological assessment. *Annual Review of Psychology*, *45*(1), 1–23. doi:10.1146/annurev.ps.45.020194.000245 PMID:8135502

Bigler, E. D. (2007). A motion to exclude and the 'fixed' versus 'flexible' battery in 'forensic' neuropsychology: Challenges to the practice of clinical neuropsychology. *Archives of Clinical Neuropsychology*, *22*(1), 45–51. doi:10.1016/j.acn.2006.06.019 PMID:17194563

Bruce Goldstein, E. (2002). *Sensation and Perception*. Pacific Grove, CA: Wadsworth.

Golden, C. J. (Eds.). (1983). Foundations of Clinical Neuropsychology. Plenum Press.

Golden, C. J., Purisch, A. D., & Hammeke, T. A. (1979). *The Luria-Nebraska Neuropsychological Battery: A manual for clinical and experimental uses*. Lincoln, NE: University of Nebraska Press.

Gupta, S., Khandelwal, S.K., Tandon, P.N., Maheshwari, M.C., Mehta, V.S., Sundram, K.R., … Jain, S. (2000). The Development and Standardization of a Comprehensive Neuropsychological Battery in Hindi -Adult form. *Journal of Personality and Clinical Studies*, *16*(2), 75–109.

Hunt, J. (1949). On the localization of the functions of the brain with special reference faculty of language (Part II). *Anthropological Review*, *7*(24), 100–116. doi:10.2307/3025013

Kaplan, E. (1988). The process approach to neuropsychological assessment. *Aphasiology*, *2*(3–4), 309–311. doi:10.1080/02687038808248930 PMID:2136065

Kar, B.R., Rao, S.L., Chandramouli, B.A., & Thennrasu, K. (2004). *NIMHANS Neuropsychological Battery for Children- Manual*. National Institute of Mental Health and Neurosciences.

Kessels, R. P. C., & Brands, A. M. A. (2009). Neuropsychological assessment. In: Biessels, Larrabee, G.J. (2008). Flexible vs. fixed batteries in forensic neuropsychological assessment: Reply to Bigler and Hom. *Archives of Clinical Neuropsychology*, *23*(7–8), 763–776.

Kessels, R. P. C., & Hendriks, M. P. H. (2016). Neuropsychological Assessment. In Encyclopedia of Mental Health (2nd ed.; vol. 3). Waltham, MA: Academic Press. doi:10.1016/B978-0-12-397045-9.00136-1

Lezak, M. D. (1976). *Neuropsychological assessment*. Oxford, UK: Oxford University Press.

Luchsinger, J.A. (Eds.). (2010). Diabetes and the Brain. Totowa, NJ: Humana Press.

Luria, A. R. (1966). *Higher cortical functions in man*. New York: Springer.

Mann, L. (1979). *On the trail of process: A historical perspective on cognitive processes and their training*. New York: Grune & Stratton.

McHenry, L. C. (1969). *Garrioson's History of Neurology*. Springfield, IL: Charles C. Thomas.

Pershad, D., & Verma, S. K. (2007). Hand Book of PGI Battery of Brain Dysfunction (PGIBBD) (2nd ed.). National Psychological Corporation.

Purisch, A. D. (2001). Misconceptions about the Luria-Nebraska Neuropsychological Battery. *NeuroRehabilitation, 16*(4), 275–280. doi:10.3233/NRE-2001-16412 PMID:11790914

Rao, S.L., Subbakrishna, D.K., & Gopukumar, K. (2004). *NIMHANS Neuropsychological Battery-Manual*. National Institute of Mental Health and Neurosciences.

Reitan, R. M. (1994). Ward Halstead's contributions to neuropsychology and the Halstead-Reitan Neuropsychological Test Battery. *Journal of Clinical Psychology, 50*(1), 47–70. doi:10.1002/1097-4679(199401)50:1<47::AID-JCLP2270500106>3.0.CO;2-X PMID:8150995

Reitan, R. M., & Davidson, L. D. (1974). *Clinical neuropsychology: Current status and applications*. Washington, DC: Winston.

Vaishali, H. P. (2019). Assessment of organicity in India. *International Journal of Research and Analytical Reviews, 6*(2), 322-26.

Vanderploeg, R. D. (Ed.). (2000). *Clinician's Guide to Neuropsychological Assessment* (2nd ed.). Hillsdale, NJ: Lawrence Erlbaum.

KEY TERMS AND DEFINITIONS

Double Dissociation: If one can demonstrate that a lesion in brain structure A impairs function X but not Y, and further demonstrate that a lesion to brain structure B impairs function Y but spares function X, one can make more specific inferences about brain function and function localization.

Neuropsychological Functions: Use of specifically designed tasks to measure a psychological function known to be linked to a particular brain structure or pathway. These are used for research into brain function and in a clinical setting for the diagnosis of deficits.

Test Interpretation: Neuropsychologists extract accurate information from their assessment procedures, draw accurate inferences from that information, and accurately convey the results to their clients and referring clinician in a manner they can understand.

Chapter 13
Neuropsychological Rehabilitation

Geeta Singh
All India Institute of Medical Sciences (AIIMS), India

G. S. Kaloiya
All India Institute of Medical Sciences (AIIMS), India

ABSTRACT

Neuropsychological rehabilitation is concerned with achievement of maximum potential in diverse domains of psychological, social, vocational, and everyday life functioning in people with cognitive, emotional, psychosocial, and behavioral deficits caused by an insult to the brain. Hence, neuropsychological rehabilitation is the process of helping a patient to recover the functions that are impaired due to brain injury. The major role of neuropsychological rehabilitation is the improvement of cognitive deficits resulting from brain damage using extensive cognitive retraining. This chapter explores neuropsychological rehabilitation.

CHAPTER OUTLINE

- Introduction
- Use of rehabilitation
- Rehabilitation team
- Brain plasticity
- Basics of neuropsychological rehabilitation
- Principles of rehabilitation
- categories of neuropsychological rehabilitation assessment

DOI: 10.4018/978-1-7998-2860-0.ch013

- Techniques of cognitive rehabilitation
- Cognitive rehabilitation for specific domains
- Types of cognitive retraining
- Domain wise specific methods of rehabilitation
- Development of new technologies
- Conclusion

INTRODUCTION

Neuropsychological rehabilitation is concerned with achievement of maximum potential in diverse domains of psychological, social, vocational and everyday life functioning in people with cognitive, emotional, psychosocial, and behavioral deficits caused by an insult to the brain. Hence neuropsychological rehabilitation is the process of helping a patient to recover the functions which are impaired due to brain injury. The major role of neuropsychological rehabilitation is the improvement of cognitive deficits resulting from brain damage using extensive cognitive retraining. In addition, it may also incorporate counseling and other techniques for augmenting the recovery process of the patient (Rao, 1991).

USE OF REHABILITATION

Neuropsychological rehabilitation is used for people with various conditions such as:

- Traumatic Brain Injury (TBI)
- Stroke
- Tumor
- Epilepsy
- Degenerative conditions
- Alzheimer's disease,
- Schizophrenia
- Attention-deficit disorder
- Learning disabilities

REHABILITATION TEAM

The rehabilitation team involves rehabilitation specialists consisting of nurses, psychiatrists, social worker, psychologist, speech therapists, physiotherapists, occupational therapists and others. Hence it involves multidisciplinary approach and each disciple works in coordination with each other to attain specific treatment goals.

BRAIN PLASTICITY

Neuropsychological rehabilitation is based on plasticity of the brain. Plasticity is known as the brain's capacity to adapt to changes in the environment. In severe brain injury, plasticity is the capacity of one part of brain to take on the function of a damaged part of the brain. Since adaptive function is the integrated function of brain, cognitive rehabilitation considers a thorough evaluation of the individual's profile of cognitive abilities, personality, emotional make-up, motivation, and social situation and most importantly, his/her cognitive reserve to formulate any intervention program. Recovery and restitution also depends upon the depth, size, and site of the lesion and severity of the injury. Evidence of brain plasticity is based on three functions of brain which are regeneration, reorganization, and recovery.

1. **Regeneration:** Nerve regeneration is the reconstruction and renewal of cell structure and function which is generally restricted to myelinated nerve fibers (Medical Dictionary, 2008). *Regeneration* occurs primarily through 'collateral sprouting' (Liu and Chambers, 1958). In catecholaminergic fibres, during axonal recovery, growing zones tend to invade vacant terminal spaces; and neighboring neurons also may sprout and send fibres if possible. Collateral sprouting occurs both from intact axons and from the collaterals of damaged ones. Some central structure responds more to stimulation after brain injury. However, the remaining fibers from the damaged area may produce a greater effect on the denervated region, promoting recovery.
2. **Reorganization:** 'Reorganization' is basically related to the neural circuits. It is about strengthening of existing neural connection and also the establishment of new connections by collateral sprouting. Remediation approach and behavior modification program works at long term reorganization of these neural circuits through strengthening of the stimulus response bond.
3. **Recovery:** Recovery from brain damage due to trauma can be divided into two stages:

a. The first stage is recovery from the acute effects of metabolic and membrane failure. The recovery of brain cells' functions is known to cause rapid recovery in neurological deficit after a brain damage.
b. Second stage recovery is based on the fact that nervous system damage may lead to a variety of compensatory changes. Gradual improvement after damage could be due to the learning of new cognitive behavioral strategies rather than recovery of lost function. Kapur et al. (2008) also explained that attained improvement was not result of recovered lost cognitive function but it was because their cognition allowed them to accomplish cognitive tasks in alternative way.

Thus the evidence of brain plasticity shows that cognitive rehabilitation approach can be 'restorative' (focus: weakness) as well as 'compensatory' (focus: strength), bringing us back to organization of intact circuitry and or development of novel circuitry.

BASICS OF NEUROPSYCHOLOGICAL REHABILITATION

There are several basics of neuropsychological rehabilitation (Prigatano, 1999):

- Clinicians must understand the patient's perspective and subjective experience to deal with their frustrations and it also helps to engage them in rehabilitation process.
- The patients' symptoms not only represent neuropsychological deficits caused by organic brain pathology but it also involves their premorbid cognitive and personality characteristics.
- Neurological rehabilitation combines both the remediation of higher cognitive dysfunctions and their management in interpersonal or social situations
- Rehabilitation process also helps patients to observe their behavior and understand the direct and indirect effects of brain damage. This facilitates them to avoid destructive choices and manage their catastrophic reactions.
- It is essential to study the interaction of cognition and personality. Failure to study these interactions may cause an inadequate understanding of neuropsychological rehabilitation process and cognitive sciences.
- Neuropsychological rehabilitation also often incorporates psychotherapeutic interventions because these intervention programs may help patients and their families to deal with their other personal issues.
- It is true that dealing with the brain dysfunctional patients produces affective reactions in both the patients' family and the staff involved in the rehabilitation

process. Therefore appropriate management of these affective reactions is essential to facilitate the rehabilitative processes.

- Neuropsychological rehabilitation program is a dynamic process. There will be either development or decline. Hence, the objective and scientific investigation helps the rehabilitation staff understand their own success and failure and is facilitated to form a dynamic, creative, rehabilitation effort.
- It is important to judge which patients will be benefited or not benefited by neuropsychological rehabilitation approaches as it enhances the credibility of the field.
- Most of the time disturbances in self awareness caused by brain damage has been poorly understood and inadequately managed.
- It is needed to well read about the mechanisms of recovery and deterioration symptoms after brain injury as it helps better patient management and also to planning creatively the rehabilitation program.

PRINCIPLES OF REHABILITATION

1. **Therapeutic Milieu:** It is essential to provide an environment which is safe, trustable and cooperative. The therapeutic surrounding needs to pleasant and comfortable. Patient and staff should be working together in a collaborative manner.
2. **Understanding and Assessment of Patients' Strength and Weaknesses**: The understanding involves biopsychosocial approach to understand a person as whole. It includes understanding about patients' following conditions:
    a. **Brain Pathology:** It is important to find out the nature of brain pathology such as head injury or stroke etc. The diagnosis may help a clinician to investigate in a right direction and formulate an adequate Intervention plan.
    b. **Cognitive functions:** Comprehensive evaluation of patients' cognitive functions includes assessment of memory, perception, language, attention and executive functions
    c. **Affect:** Assessment for clinical conditions such as depression, anxiety, anger, and as well as of motivation and confidence level
    d. **Physical Issues:** Rehabilitation is always planned considering patients' medical condition. It is important to know whether patient is suffering from heiplegia, any kind of sensory loss, dysarthria or pain
    e. **Functional Consequences:** The functional limitation of the patients such as problem in their work, relationships, daily living activities etc.

f. **Family/Social Support:** It includes the details with whom patient stays, supportive network in family and interpersonal relationship issues
g. Premorbid factors including coping style, personality features etc.

Hence multi-disciplinary assessment is conducted by rehabilitation specialists and then results are integrated together for formulation of summary about the patient.

CATEGORIES OF NEUROPSYCHOLOGICAL REHABILITATION ASSESSMENT

The assessment can be categorized based on two models:

Quantitative Model: It is centered in the lesion and the cognitive deficits the patient show on standardized tests and neuropsychological batteries. The focus of this model is mainly on the disorder or deficits of the patient caused by damage of the brain.

Qualitative Model: It is based on the principles of Luria. It focuses on deficits as well as intact functions in a particular patient. The neuropsychological rehabilitation starts with what intact functions rather than what has been damaged. Thus the main objective for patients remains to achieve neuropsychological coherence and not just returning to their earlier normalcy (Leon-Carrion, 1986).

There are basically two standard comprehensive batteries that are widely used for assessment for rehabilitation: Halstead Reitan Battery (Reitan & Davison, 1974) and Luria Nebraska Neuropsychological batteries (LNNB) (Golden, Purisch, & Hammeke, 1980). These batteries have standard administration process and scoring methods that provides assessment of the various cognitive perceptual and motor abilities. These areas included general intelligence, abstraction and related conceptual abilities, attention, language, memory and a variety of perceptual motor skills.

In addition to these batteries, appropriate Weschler Intelligence Scale and its subtests depending on the need of the patient may be used. Many researchers also use Weschler Memory Scale, Boston Diagnostic Aphasia Scale. A single test widely used for neuropsychological rehabilitation is Wisconsin Card Sorting Test (Heaton, Chelune, Talley, Kay, & Curtiss, 1993). It is basically a test of executive functions that assess planning, concept formation and conceptual flexibility (Grant & Berg, 1948).

Besides these standardized behavioral observations and assessments play a vital role. Behavioral interviewing scales, checklists and direct observation can all be used to identify and monitor the everyday implication of neuropsychological impairment.

These test batteries, individual tests and behavioral assessment should fulfill the following needs:

To understand the cognitive abilities and deficits in order to find out what is cognitively possible and not possible for the patient.
How the cognitive impairment is leading to problems in day to day life and what is important for the people involved in the treatment process.
Individualize treatment such individualizations involves the identification of problems from the patient's point of view.

1. **Goal Setting:** The goal of rehabilitation is mainly facilitating patients to actively participate in work, education, social and independent living. An achievable goal is set through mutual consensus between patient and rehabilitation clinician during the course of rehabilitation program. There should be following characteristics of goal:
    a. Specific
    b. Measurable
    c. Achievable
    d. Timeframe set
    e. Relevant to patients
2. **Development of Compensatory Strategies and Retraining Skills:** Neuropsychological rehabilitation use two primary approaches:
    a. Approaches that focus on retraining of an impaired skill, such as attention or memory. These are known as cognitive remediation or cognitive retraining. Cetrain tasks as practice exercise may be given to the patient to attempt to strengthen memory or attention assuming that the brain may be able to rebuild axonal connections through retraining.
    b. Approaches that emphasize on searching for adaptations to the person's environment in the context in which the person will use them.

TECHNIQUES OF COGNITIVE REHABILITATION

Techniques of cognitive remediation fall into following three categories

1. Restoration
2. Compensatory
3. Holistic Approach

a. **Restoration:** Restoration involves regaining function by remediating the impaired cognitive ability (for example, selective attention, spatial perception, or memory retrieval) using systematic training. Engaging patients in repetitive exercises is the primary technique to ameliorate the specific cognitive deficit. Restorative approaches take a "bottom-up" approach; remediation is directed towards restoring a particular area of cognitive functioning (such as attention) rather than towards developing strategies for successfully performing specific work tasks.

b. **Compensation:** Compensation focuses on the disability (for example, following written instructions, navigating the community, or remembering appointments) and trains a person to use strategies to again accomplish these tasks. The techniques consist of training in use of external prosthetics and self-cueing for the purpose to develop strategies for accomplishing specific functions. This also includes determining the specific activities affected by the individual's cognitive impairment (for instance; organizing tasks, keeping a schedule etc).

   Restorative techniques would be used to focus on weaknesses, while compensatory techniques are often driven by a person's strengths, which serve as the basis for developing effective strategies.

c. **Holistic Approach:** In contrast to retraining and compensatory approaches that focus primarily on specific cognitive processes, holistic programs attempt cognitive rehabilitation in the context of the full range of problems experienced by brain damaged individuals. The holistic approaches are predicated on several key assumptions.

   i. Cognitive therapies must not be isolated from other (physical, psychological)
   ii. therapies that are essential to functional rehabilitation.
   iii. Cognitive rehabilitation proceeds in hierarchical way from lower level function to
   iv. more complex functions.
   v. Orderly and structured therapeutic programs are required for the patients with the
   vi. disorganized and chaotic nature of thought processes caused be brain damage.
   vii. Because the problems of brain injured people are highly personalized, treatment must
   viii. be as individualized as possible.

4. **Implementation of Psychological Intervention:** Neuropsychological rehabilitation is not only restricted to rehabilitation of cognition but also considers personality, emotional issues and awareness of an individual. Personality changes

are very common after brain injury (Jennett & Teasdale, 1981) as reported by relatives and significant others and therapists attempts to treat personality related issues as it complicates the rehabilitation process. Psychotherapy helps in building motivation of patients and deal with their frustration in therapy process. Patients with brain injury often report declined social life (Elsass & Kinsella, 1987) and loneliness as frequent complaint (Thomsen, 1974). Poor social interaction is often evident among these patients (Newton & Johnson, 1985).

Frontal lobe damage is typically associated with impulsivity, lack of initiation, disinhibition, rigidity, poor social judgment, loss of abstract attitude, and loss of personal and social awareness (Lezak, Howieson, Loring, Hannay, & Fischer, (2004). Prigatano (1999) states that self-awareness is the product of integration of "thought" and "feeling" areas of the brain by combining inputs from sensorimotor, limbic, and paralimbic areas. Taking perspective of others varies based on severity of brain injury (Spiers, Pouk, & Santoro, 1994; Santoro & Spiers, 1994) and therefore may lead to apparent egocentric or unempathic behavior. The greater awareness of own dysfunction typically show higher levels of emotional distress (Nockleby & Deaton, 1987).

Hence psychotherapy in the rehabilitation process often addresses neuropsychology-based issues of motivation, self-awareness, egocentricity, emotionality, personality and empathy. Psychotherapy can be helpful to aid in effective coping and adjustment for a patient. Inadequate psychosocial functioning and poor self awareness often are the primary reasons for vocational difficulties. Research indicates that dealing with emotional issues could help making another 20% to 30% of patients productive workers (Prigatano, 1999). It is also important to involve the family members and care givers in the rehabilitation process.

COGNITIVE REHABILITATION FOR SPECIFIC DOMAINS

Cognitive Rehabilitation is a wide term which involves variety of intervention strategies such as:

1. **Cognitive Retraining:** Retraining programs usually involve repeated practice of specific cognitive exercises designed to strengthen basic skills (attention, encoding) that are essential for more complex cognitive function. Cognitive retraining is a promising adaptation of scaffolding approach whereby training begins with basic skills and further progresses to complex skills. This approach is clearly reflected in Sohlberg and Mateer's (1989) 'process-oriented' model in

which component processes that are presumed to mediate a particular cognitive skill are trained in a hierarchical order, such that progressively more complex processes are trained as learning occurs . Within each level of training, skills are repetitively practiced, under the assumption that repeated practice in a structured setting is a necessary component to strengthen and reautomatise cognitive skills. Cognitive retraining should be embedded in real life, everyday contexts, building on residual or already learnt cognitive routines. Tasks may also be personalized to some degree. Regular feedback must be given.

2. **Compensatory Strategy Training:** Compensatory approaches are based on the idea that reduced cognitive abilities can be compensated by making use of specific techniques that can help in organizing information while learning or retrieval process.

   Compensatory strategies include:

   - **Internal strategies:** Internal strategies can help individuals to make the necessary effort to solve a problem. Such techniques place considerable demands on the individual's ability to use them appropriately but, as West (1995) pointed out, an added benefit of this effort is increased attention to the task at hand.
   - **External Strategies:** With external strategies, the objective is to identify distinctive cues that can be readily associated with to-be-remembered material. External strategies focus on use of objects present in the surrounding to compensate for the cognitive deficits.
   - **Domain – Specific Techniques:** Domain specific approaches capitalize on preserved domain or systems to train functions that are normally mediated by other impaired system.

TYPES OF COGNITIVE RETRAINING

There are seven types of cognitive retraining approaches were illustrated in following order:

1. Attention and concentration retraining
2. Memory retraining
3. Organizational skills retraining
4. Reasoning
5. Problem solving
6. Decision making
7. Executive skills.

Attention Retraining: Attention retraining helps to improve several abilities such as focused attention, divided attention, sustain attention and the same time reducing the feeling of boredom, fatigue and distraction. The retraining reduces the time pressure and improves stimulation by alteration in the environment. It is necessary that individual should have realistic expectations regarding the attention and self-regulation abilities. Smaller tasks, more reinforcement, and alternating tasks of varying interest are essential components of attention retraining. Retraining involves usage of job cards, checklists, outlines, colour coding, graphic organizers and repeating instructions. Attention process training involves creating tasks ("exercises") that engage attention functions in a hierarchical manner, which is then, practiced using massed (back-to-back) learning trials.

One of the techniques developed by Sohlberg and Mateer (1987,1989) intended to improve deficits in four areas of attention. The basic idea behind the repetition in the targeted area is to provide repeated stimulation to the brain's attention system; which is considered to facilitate strengthening and repair of the neuronal system by restitution process of function.

The four areas include:

1. Sustained attention- to train sustained attention or hierarchies of tasks that require constant attention presented.
2. Selective attention training- incorporates distraction into tasks such as those used for sustained attention.
3. Alternative attention training- It entails switching between two attention requiring tasks.
4. Divided attention raining- It entails simultaneous attention to two or more tasks. The training may be conducted by making use of first training performance on two single sustained attention tasks.

Memory Retraining: In memory training, compensatory approach such as maintaining a daily log or schedule to using voice-activated dictation devices is made. Strategies for memory such as association, chunking, and the use of mnemonics that can be trained in therapy and then generalized to functional settings are used. Among the core processes of memory (encoding, storage, retrieval), encoding and retrieval can be significantly impaired. However, storage is often relatively spared.

Strategies for managing memories

- Using a Dictaphone/ tape recorder to record meetings, lectures ie as an aid to memory
- Using a watch with an alarm or a paging device as a cue to look at diary or to perform
- particular task .This may also increase immediate arousal level sufficiently to overcome initiation impairments
- Systematically using written or photo journals for episodic memory deficits (memory for events)
- Using notes, calendars and list as external cues and reminders. These may also helps with initiation difficulty.
- Introducing a notice board or white board in a prominent place in the home to prompt
- patients about specific actions as well as daily / weekly schedule
- Using errorless learning when teaching new skills to minimize the need to unlearn mistakes that is promptings and cueing in such a way that no errors are made during the training process.

Developing internal memory strategies through five steps, those are

- Repetition
- Association
- Chunking
- Maximizing the relevance and depth of understanding of material.
- Mnemonics:
    - Pegword method
    - Method of loci
    - Image name match method

Reasoning: Reasoning retraining techniques include listing the facts or reality of a situation; excluding irrelevant facts or details; putting the steps to solve a problem in a logical order, avoiding irrational thinking, such as jumping to conclusions based on incomplete information, focusing on the negative aspects of the situation and ignoring the positive. It is essential for a person to connect relevant information in a logical way as it enhances their comprehension level.

Problem Solving: Problem-solving retraining helps an individual to define a problem and find out with possible solutions to it. The solutions are discussed with others and advices are taken to review the various possible solutions from number of perspectives. At last it is evaluated whether the problem was solved or not. These steps be repeated several times until the problem is solved. This process is referred to as "SOLVE," from the first letter of the name of each step: Specify; Options; Listen; Vary and evaluate. The technique is more appropriate to use with individuals with higher level of functioning. Providing problem solving skills involves, training to help develop an explicit and systematic approach to solving difficulties.

Decision-Making Retraining: Decision-making retraining is utilized when a person need to choose among a number of options. The retraining aims to help an individual to consider their decisions thoroughly before taking any action. The considerations may range from such practical matters as money, people, rules and policies, to personality issues.

Executive Skills: Executive skills retraining refers to teaching individuals how to monitor themselves, control their thinking and actions, think in advance, set goals, manage time, act in socially acceptable ways, and transfer skills to new situations. These are higher-level cognitive skills. Retraining of executive skills makes use of charts and videotapes for monitoring behavior with use of a variety of questions, tasks, and games.

Both restorative and compensatory strategies are used for the remediation.

Compensatory strategies are used if the patient is very acute and is demonstrating very severe executive function disturbance, focus should be on teaching task specific routines. They may profit from training of particular sequences for standard highly repetitive functional activities such as showering, taking care of one's toilet or dressing. Behavior techniques include shaping and reinforcement.

DOMAIN WISE SPECIFIC METHODS OF REHABILITATION

Attention

- **Digit Cancellation Task:** Particular number will be cancelled from a list of random numbers.
  There was a gradual increase in the number of digits, reduction of size of digits and increased scatter of the digits to increase the difficulty level.
- **Number Connection:** Numbers placed on blank paper will be connected serially. Total numbers to connect will be increased to enhance the task difficulty

Mental Speed

- **Digit Coding Test:** Corresponding digit will be substituted by designated symbol in a specified time period.
- **Grain Sorting Test:** Sorting of mixed grains will be done by piling of similar grains. Task difficulty will be increased by increasing the type of grains.

Working Memory

Sequencing of jumbled letters: Letters will be rearranged to make a meaningful word which will have an average of 4-6 letters.

Visuo-Constructive Ability

Dot Based Design: Three geometrical designs will be constructed using the dots placed in each of the 10 boxes

Planning

- **Maze Completion:** The way to go out will be traced in blind alley beginning with the start point. Task difficulty will be increased by increasing the complexity of maze.
- **Short Essay Writing:** Short essay writing will be done on given topic considering the patient's interest and context.

DEVELOPMENT OF NEW TECHNOLOGIES

- Computer-Assisted Cognitive Rehabilitation (CACR)
- Graphic advance organisers
- Neuro-feedback
- Virtual reality
- Prosthetic aids

CONCLUSION

Neuropsychological rehabilitation is an effective mode of treatment in a number of conditions including organic and other psychiatric conditions. The result is the emergence of the holistic approaches, which have been associated with positive outcomes. Neuropsychological rehabilitation plays an important role in evaluation of the cognitive profile of patients with various neurological conditions. It aims to know that how such condition lead to specific cognitive deficits and dysfunction in day to day life. The rehabilitation process utilizes creative ways to deal with deficits of individuals and to help them to ameliorate dysfunction caused by brain damage.

REFERENCES

Elsass, L., & Kinsella, G. (1987). Social interaction following severe closed head injury. *Psychological Medicine*, *17*(1), 67–78. doi:10.1017/S003329170001299X PMID:3575579

Golden, C. J., Purisch, A. D., & Hammeke, T. A. (1980). *The Luria-Nebraska Neuropsychological Battery: Manual (Revised)*. Los Angeles, CA: Western Psychological Services.

Grant, D. A., & Berg, E. (1948). A behavioral analysis of degree of reinforcement and ease of shifting to new responses in a Weigl-type card-sorting problem. *Journal of Experimental Psychology*, *38*(4), 404–411. doi:10.1037/h0059831 PMID:18874598

Heaton, R. K., Chelune, G. J., Talley, J. L., Kay, G. G., & Curtiss, G. (1993). *Wisconsin Card Sorting Test Manual: Revised and Expanded*. Odessa, FL: Psychological Assessment Resources.

Jennett, B., & Teasdale, G. (1981). *Management of head injuries*. Philadelphia: F. A. Davis Company.

Kapur, S., Craik, F. I. M., Tulving, E., Wilson, A. A., Houle, S., & Brown, G. M. (1994). Neuroanatomical correlates of encoding in episodic memory: Levels of processing effect. *Proceedings of the National Academy of Sciences of the United States of America*, *91*(6), 2008–2011. doi:10.1073/pnas.91.6.2008 PMID:8134340

Leon-Carrion. (1986). *Diagnostico Clinico en Psicologia*. Sevilla: Alfar.

Lezak, M. D., Howieson, D. B., Loring, D. W., Hannay, H. J., & Fischer, J. S. (2004). *Neuropsychological assessment* (4th ed.). Oxford University Press.

Liu, C. N., & Chambers, W. W. (1958). Intraspinal sprouting of dorsal root axons. *Archives of Neurology*, *79*, 46–61. PMID:13486981

Medical Dictionary. (2020). Retrieved from https://medical-dictionary.thefreedictionary.com/regeneration

Newton & Johnson. (1985). Principles of Neuropsychology (2nd ed.). Academic Press.

Nockleby, D. M., & Deaton, A. V. (1987). Denial versus distress: Coping patterns in post head trauma patients. *The International Journal of Clinical Neuropsychology*, *9*(4), 145–148.

Prigatano, G. P. (1999). *Principles of Neuropsychological Rehabilitation*. Oxford University Press.

Prigatano, G. P., & Leathem, J. M. (1993). Awareness of behavioral limitations after traumatic brain injury: A cross-cultural study of New Zealand Maoris and non-Maoris. *Clinical Neuropsychologist*, *7*(2), 123–135. doi:10.1080/13854049308401514 PMID:29022474

Reitan, R. M., & Davison, L. A. (Eds.). (1974). *Clinical neuropsychology: Current status and applications*. V. H. Winston & Sons.

Santoro, J., & Spiers, M. (1994). Social cognitive factors in brain injury-associated personality change. *Brain Injury: [BI]*, *8*(3), 265–276. doi:10.3109/02699059409150978 PMID:8004084

Sohlberg, M. M., & Mateer, C. A. (1987). Effectiveness of an attention training program. *Journal of Clinical and Experimental Neuropsychology*, *19*(2), 117–130. doi:10.1080/01688638708405352 PMID:3558744

Spiers, M. V., Pouk, J. A., & Santoro, J. M. (1994). Examining perspective-taking in the severely head-injured. *Brain Injury: [BI]*, *8*(5), 463–473. doi:10.3109/02699059409150998 PMID:7951209

Thomsen, I. V. (1974). The patient with severe head injury and his family. *Scandinavian Journal of Rehabilitation Medicine*, *6*, 180–183. PMID:4458060

West, R. L. (1995). Compensatory strategies for age associated memory impairment. In Hand Book of Memory Disorders (pp. 481-500). Wiley.

KEY TERMS AND DEFINITIONS

Alzheimer's Disease: Is an irreversible, progressive brain disorder that slowly destroys memory and thinking skills and, eventually, the ability to carry out the simplest tasks. In most people with the disease those with the late-onset type-symptoms first appear in their mid-60s.

Attention Deficit Disorder: Is a neurological disorder that causes a range of behavior problems such as difficulty attending to instruction, focusing on schoolwork, keeping up with assignments, following instructions, completing tasks and social interaction.

Degenerative Conditions: Is a type of a medical condition that causes a tissue or organ to deteriorate over time. There are quite a number of degenerative diseases and many of them are associated with aging or gets worse during the aging process.

Epilepsy: Is a central nervous system (neurological) disorder in which brain activity becomes abnormal, causing seizures or periods of unusual behavior, sensations, and sometimes loss of awareness. Anyone can develop epilepsy.

Learning Disabilities: Learning disabilities are neurologically-based processing problems. These processing problems can interfere with learning basic skills such as reading, writing and/or math. They can also interfere with higher level skills such as organization, time planning, abstract reasoning, long or short-term memory and attention. It is important to realize that learning disabilities can affect an individual's life beyond academics and can impact relationships with family, friends and in the workplace.

Schizophrenia: Is a chronic and severe mental disorder that affects how a person thinks, feels, and behaves. People with schizophrenia may seem like they have lost touch with reality. Although schizophrenia is not as common as other mental disorders, the symptoms can be very disabling.

Stroke: Is a medical condition in which poor blood flow to the brain results in cell death. There are two main types of stroke: ischemic, due to lack of blood flow, and hemorrhagic, due to bleeding. Both result in parts of the brain not functioning properly.

Traumatic Brain Injury (TBI): Traumatic brain injury (TBI) is sudden damage to the brain caused by a blow or jolt to the head. Common causes include car or motorcycle crashes, falls, sports injuries, and assaults. Injuries can range from mild concussions to severe permanent brain damage.

Tumor: Is a mass of tissue that's formed by an accumulation of abnormal cells. Normally, the cells in your body age, die, and are replaced by new cells. With cancer and other tumors, something disrupts this cycle. Tumor cells grow, even though the body does not need them, and unlike normal old cells, they don't die.

Compilation of References

Abib, R. T., Gaman, A., Dargél, A. A., Tamouza, R., Kapczinski, F., Gottfried, C., & Leboyer, M. (2018). Intracellular Pathogen Infections and Immune Response in Autism. *Neuroimmunomodulation*, *25*(5-6), 271–279. doi:10.1159/000491821 PMID:30130799

Agarwal, N., Port, J. D., Bazzocchi, M., & Renshaw, P. F. (2010). Update on the use of MR for assessment and diagnosis of psychiatric diseases. *Radiology*, *255*(1), 23–41. doi:10.1148/radiol.09090339 PMID:20308442

Akanuma, N., Alarcón, G., Lum, F., Kissani, N., Koutroumanidis, M., Adachi, N., ... Morris, R. G. (2003). Lateralising value of neuropsychological protocols for presurgical assessment of temporal lobe epilepsy. *Epilepsia*, *44*(3), 408–418. doi:10.1046/j.1528-1157.2003.24502.x PMID:12614397

Albert, D. J., Walsh, M. L., & Jonik, R. H. (1993). Aggression in humans: What is its biological foundation? *Neuroscience and Biobehavioral Reviews*, *17*(4), 405–425. doi:10.1016/S0149-7634(05)80117-4 PMID:8309650

Allen, C. C., & Ruff, R. M. (1990). Self-rating versus neuropsychological performance of moderate versus severe head-injured patients. *Brain Injury*, *4*(1), 7–17. doi:10.3109/02699059009026143 PMID:2297602

Amadio, J. P., Beasley, K., Rothstein, B. D., Selman, W. R., Riley, J., & Emery, J. (2015). Intervening in the Brain: Ethics and Neurosurgery. *The Virtual Mentor*, *17*, 1.

American Psychiatric Association. (2013). *Diagnostic and Statistical Manual of Mental Disorders* (5th ed.). Arlington, VA: American Psychiatric Association.

Aversi-Ferreira, T. A., Tamaishi-Watanabe, B. H., Magri, M. P. D. F., & Aversi-Ferreira, R. A. (2019). Neuropsychology of the temporal lobe: Luria's and contemporary conceptions. *Dementia & Neuropsychologia*, *13*(3), 251–258. doi:10.1590/1980-57642018dn13-030001 PMID:31555397

Bakker, A., Albert, M. S., Krauss, G., Speck, C. L., & Gallagher, M. (2015). Response of the medial temporal lobe network in amnestic mild cognitive impairment to therapeutic intervention assessed by fMRI and memory task performance. *NeuroImage. Clinical*, *7*, 688–698. doi:10.1016/j.nicl.2015.02.009 PMID:25844322

Barnett, A. J., Park, M. T. M., Pipitone, J., Chakravarty, M. M., & McAndrews, M. P. (2015). Functional and structural correlates of memory in patients with mesial temporal lobe epilepsy. *Frontiers in Neurology*, *6*, 103. doi:10.3389/fneur.2015.00103 PMID:26029159

Barthas, F., & Kwan, A. C. (2017). Secondary motor cortex: Where 'sensory'meets 'motor'in the rodent frontal cortex. *Trends in Neurosciences*, *40*(3), 181–193. doi:10.1016/j.tins.2016.11.006 PMID:28012708

Beaumont, J. G. (1983). *Introduction to neuropsychology*. Oxford, UK: Blackwell Scientific Publications.

Beaumont, J. G. (2008). *Introduction to Neuropsychology* (2nd ed.). The Guilford Press.

Bengesser, S. A., Mörkl, S., Painold, A., Dalkner, N., Birner, A., Fellendorf, F. T., ... Reininghaus, E. Z. (2019). Epigenetics of the molecular clock and bacterial diversity in bipolar disorder. *Psychoneuroendocrinology*, *1*(101), 160–166. doi:10.1016/j.psyneuen.2018.11.009 PMID:30465968

Bennett, M. V., & Zukin, R. S. (2004). Electrical coupling and neuronal synchronization in the mammalian brain. *Neuron*, *41*(4), 495–511. doi:10.1016/S0896-6273(04)00043-1 PMID:14980200

Benton, A. L. (1994). Neuropsychological assessment. *Annual Review of Psychology*, *45*(1), 1–23. doi:10.1146/annurev.ps.45.020194.000245 PMID:8135502

Bigler, E. D. (2007). A motion to exclude and the 'fixed' versus 'flexible' battery in 'forensic' neuropsychology: Challenges to the practice of clinical neuropsychology. *Archives of Clinical Neuropsychology*, *22*(1), 45–51. doi:10.1016/j.acn.2006.06.019 PMID:17194563

Binder, J. R. (2017). Current controversies on Wernicke's area and its role in language. *Current Neurology and Neuroscience Reports*, *17*(8), 58. doi:10.100711910-017-0764-8 PMID:28656532

Bisiach, E., Vallar, G., Perani, D., Papagno, C., & Berti, A. (1986). Unawareness of disease following lesions of the right hemisphere: Anosognosia for hemiplegia and anosognosia for hemianopia. *Neuropsychologia*, *24*(4), 471–482. doi:10.1016/0028-3932(86)90092-8 PMID:3774133

Bittigau, P., & Ikonomidou, C. (1997). Glutamate in neurologic diseases. *Journal of Child Neurology*, *12*(8), 471–485. doi:10.1177/088307389701200802 PMID:9430311

Bonelli, S. B., Powell, R. H., Yogarajah, M., Samson, R. S., Symms, M. R., Thompson, P. J., ... Duncan, J. S. (2010). Imaging memory in temporal lobe epilepsy: Predicting the effects of temporal lobe resection. *Brain*, *133*(4), 1186–1199. doi:10.1093/brain/awq006 PMID:20157009

Booij, L., van der Does, A. J., Haffmans, P. M., Spinhoven, P., & McNally, R. J. (2005). Acute tryptophan depletion as a model of depressive relapse: Behavioural specificity and ethical considerations. *The British Journal of Psychiatry*, *187*(2), 148–154. doi:10.1192/bjp.187.2.148 PMID:16055826

Boone, K. B., Miller, B. L., Rosenberg, L., Durazo, A., McIntyre, H., & Weil, M. (1988). Neuropsychological and behavioral abnormalities in an adolescent with frontal lobe seizures. *Neurology, 38*(4), 583–586. doi:10.1212/WNL.38.4.583 PMID:3352916

Braun, C. M., Delisle, J., Guimond, A., & Daigneault, R. (2009). Post unilateral lesion response biases modulate memory: Crossed double dissociation of hemispheric specialisations. *Laterality, 14*(2), 12264. doi:10.1080/13576500802328613 PMID:18991140

Bray, N. (2017). Cerebellum: The little learning brain. *Nature Reviews. Neuroscience, 18*(5), 263. doi:10.1038/nrn.2017.47 PMID:28381834

Brazis, P. W., Masdeu, J. C., & Biller, J. (2007). *Localization in clinical neurology*. Philadelphia: Lippincott Williams & Wilkins.

Breedlove, S. M., Rosenzweig, M. R., & Watson, N. V. (2007). *Biological psychology: An introduction to behavioral, cognitive, and clinical neuroscience* (5th ed.). Sinauer Associates.

Bruce Goldstein, E. (2002). *Sensation and Perception*. Pacific Grove, CA: Wadsworth.

Bryan, S. T. (2008). *The Body and Society: Explorations in Social Theory* (3rd ed.). Sage Publications.

Bui, T., & Nguyen, J. (2019). Neuroanatomy, Cerebral Hemisphere. StatPearls Publishing.

Bunevicius, A., Deltuva, V. P., Deltuviene, D., Tamasauskas, A., & Bunevicius, R. (2008). Brain lesions manifesting as psychiatric disorders: Eight cases. *CNS Spectrums, 13*(11), 950–958. doi:10.1017/S1092852900014000 PMID:19037181

Burghardt, G. M. (2009). Ethics and Animal Consciousness: How Rubber the Ethical Ruler? *The Society for the Psychological Study of Social Issues, 65*(3), 499–521. doi:10.1111/j.1540-4560.2009.01611.x

Caminero, F., & Cascella, M. (2019). Neuroanatomy, Mesencephalon Midbrain. StatPearls Publishing.

Campbell, W. W., DeJong, R. N., & Haerer, A. F. (2005). *DeJong's the neurologic examination: Incorporating the fundamentals of neuroanatomy and neurophysiology*. Philadelphia, PA: Lippincott Williams & Wilkins.

Carlson. (2005). *Foundations of Physiological Psychology*. Allyn and Bacon.

Carlson, N. R. (2008). *Foundations of physiological psychology*. Boston, MA: Allyn and Bacon.

Catafau, A. M. (2001). Brain SPECT in clinical practice. Part I: Perfusion. *Journal of Nuclear Medicine, 42*, 259–271. PMID:11216525

Chaudhary, K., Kumaran, S. S., Chandra, S. P., Wadhawan, A. N., & Tripathi, M. (2014). Mapping of cognitive functions in chronic intractable epilepsy: Role of fMRI. *The Indian Journal of Radiology & Imaging, 24*(1), 51. doi:10.4103/0971-3026.130694 PMID:24851005

Cherry, S. R., & Phelps, M. E. (1996). Imaging brain function with positron emission tomography. In A. W. Toga & J. C. Mazziotta (Eds.), *Brain Mapping—The Methods* (pp. 191–222). San Diego, CA: Academic Press.

Chisholm, H. (Ed.). (1911). *Parallelism, Psychophysical. In Encyclopædia Britannica* (11th ed.; vol. 20, p. 762). Cambridge University Press.

Christiane, S. H., Hiroshi, M., & Mario, M. (2017). *GABA and Glutamate: Their Transmitter Role in the CNS and Pancreatic Islets, GABA And Glutamate - New Developments In Neurotransmission Research.* IntechOpen. Available from: https://www.intechopen.com/books/gaba-and-glutamate-new-developments-in-neurotransmission-research/gaba-and-glutamate-their-transmitter-role-in-the-cns-and-pancreatic-islets

Churchland, P. (2007). Neurophilosophy at Work. Cambridge University Press. doi:10.1017/CBO9780511498435

Churchland, P. S. (1989). Neurophilosophy: Toward a Unified Science of the Mind–Brain. In Computational Models of Cognition and Perception. MIT Press.

Churchland, P. S. (2002). Brain-Wise: Studies in Neurophilosophy. Bradford Books.

Cloninger, C. R., Svrakic, D. M., & Przybeck, T. R. (1993). A psychobiological model of temperament and character. *Archives of General Psychiatry, 50*(12), 975–990. doi:10.1001/archpsyc.1993.01820240059008 PMID:8250684

Clower, W.T., & Finger, S. (2001). Discovering trepanation: the contribution of Paul Broca. *Neurosurgery, 49*(6), 1417-26.

Cole, M. (2017). *Selected Writings of AR Luria.* Routledge. doi:10.4324/9781315171913

Critchley, M. (1953). *The parietal lobes.* Williams and Wilkins.

Damasio, A. R. (1994). *Descartes' error: emotion, reason, and the human brain.* New York: Grosset/Putnam.

Daroff, R. B., & Bradley, W. G. (2012). *Bradley's neurology in clinical practice.* Philadelphia, PA: Elsevier/Saunders.

Dawson, G. D. (1947). Investigations on a patient subject to myoclonic seizures after sensory stimulation. *Journal of Neurology, Neurosurgery, and Psychiatry, 10*(4), 141–162. doi:10.1136/jnnp.10.4.141 PMID:18905644

Dehaene, S., Spelke, E., Pinel, P., Stanescu, R., & Tsivkin, S. (1999). Sources of mathematical thinking: Behavioral and brain-imaging evidence. *Science, 284*(5416), 9704. doi:10.1126cience.284.5416.970

Dennett, D. C. (1986). Content and Consciousness. International Library of Philosophy. Taylor & Francis.

Dennett, D. C. (1997). Kinds of Minds: Toward an Understanding of Consciousness. Science Masters Series. Basic Books.

Devinsky, O. (2009). Delusional misidentifications and duplications: Right brain lesions, left braindelusions. *Neurology*, *72*(1), 807. doi:10.1212/01.wnl.0000338625.47892.74 PMID:19122035

Ellis, J. R., Ellis, K. A., Bartholomeusz, C. F., Harrison, B. J., Wesnes, K. A., Erskine, F. F., ... Nathan, P. J. (2006). Muscarinic and nicotinic receptors synergistically modulate working memory and attention in humans. *The International Journal of Neuropsychopharmacology*, *9*(02), 175–189. doi:10.1017/S1461145705005407 PMID:15877932

Elsass, L., & Kinsella, G. (1987). Social interaction following severe closed head injury. *Psychological Medicine*, *17*(1), 67–78. doi:10.1017/S003329170001299X PMID:3575579

Everitt, B. J., & Robbins, T. W. (2005). Neural systems of reinforcement for drug addiction: From actions to habits to compulsion. *Nature Neuroscience*, *8*(11), 1481–1489. doi:10.1038/nn1579 PMID:16251991

Fields, R. D., & Stevens-Graham, B. (2002). New insights into neuron-glia communication. *Science*, *298*(5593), 556–562. doi:10.1126cience.298.5593.556 PMID:12386325

Fischbach, G. D. (2006). Schizophrenia: Signals from the other side. *Nature Medicine*, *12*(7), 734–735. doi:10.1038/nm0706-734 PMID:16829916

Fischer, S., Trexler, L. E., & Gauggel, S. (2004b). Awareness of activity limitations and prediction of performance in patients with brain injuries and orthopedic disorders. *Journal of the International Neuropsychological Society*, *10*(2), 190–199. doi:10.1017/S1355617704102051 PMID:15012839

Fleming, J. M., Strong, J., & Ashton, R. (1996). Cluster analysis of self-awareness levels in adults with traumatic brain injury and relationshipto outcome. *The Journal of Head Trauma Rehabilitation*, *13*(5), 39–51. doi:10.1097/00001199-199810000-00006 PMID:9753534

Folk, G. E. Jr, & Long, J. P. (1988). Serotonin as a neurotransmitter: a review. *Comparative Biochemistry and Physiology Part C Toxicology & Pharmacology*, *91*(1), 251-7.

Frank, W. P. (2009). Roles of glial cells in synapse development. *Cellular and Molecular Life Sciences*, *66*(13), 2037–2047. doi:10.100700018-009-0005-7 PMID:19308323

Friederici, A. D. (2018). The neural basis for human syntax: Broca's area and beyond. *Current Opinion in Behavioral Sciences*, *21*, 88–92. doi:10.1016/j.cobeha.2018.03.004

Gallagher, S. (2005). How the Body Shapes the Mind. New York: Oxford University Press. doi:10.1093/0199271941.001.0001

Gelder, M. J. (1996). Biological Psychiatry in perspective. *British Medical Bulletin*, *52*(3), 401–407. doi:10.1093/oxfordjournals.bmb.a011556 PMID:8949246

Gibson, W. C. (1962). Pioneers of localization of function in the brain. *Journal of the American Medical Association*, *180*, 944–951. PMID:13898444

Gittins, R., & Harrison, P.J. (2004). Neuronal density, size and shape in the human anterior cingulate cortex: a comparison of Nissl and NeuN staining. *Brain Research Bulletin, 63*(2), 155-60.

Goldberg, E. (Ed.). (2019). *Contemporary neuropsychology and the legacy of Luria*. Psychology Press. doi:10.4324/9780203771662

Golden, C. J. (Eds.). (1983). Foundations of Clinical Neuropsychology. Plenum Press.

Golden, C. J., Purisch, A. D., & Hammeke, T. A. (1979). *The Luria-Nebraska Neuropsychological Battery: A manual for clinical and experimental uses*. Lincoln, NE: University of Nebraska Press.

Golden, C. J., Purisch, A. D., & Hammeke, T. A. (1980). *The Luria-Nebraska Neuropsychological Battery: Manual (Revised)*. Los Angeles, CA: Western Psychological Services.

Golgi, C. (1873). Sulla sostanzagrigiadelcervello. *GazettaMedicalItaliana., 33*, 244–246.

Grafman, J., Vance, S. C., Weingartner, H., Salazar, A. M., & Amin, D. (1986). The effects of lateralized frontal lesions on mood regulation. *Brain, 109*(6), 1127–1148. doi:10.1093/brain/109.6.1127 PMID:3790971

Grant, D. A., & Berg, E. (1948). A behavioral analysis of degree of reinforcement and ease of shifting to new responses in a Weigl-type card-sorting problem. *Journal of Experimental Psychology, 38*(4), 404–411. doi:10.1037/h0059831 PMID:18874598

Graziane & Sweet. (2017). Comprehensive textbook of psychiatry (10th ed.). Wolters Kluwer.

Grem, J.L,, King, S.A., O'Dwyer, P.J., & Leyland-Jones, B. (1988). Biochemistry and clinical activity of N-(phosphonacetyl)-L-aspartate: a review. *Cancer Research, 48*(16), 4441-54.

Gross, L. (2006). Supporting" players take the lead in protecting the overstimulated brain. *PLoS Biology, 4*(11).

Guloksuz, S., & Van Os, J. (2018). The slow death of the concept of schizophrenia and the painful birth of the psychosis spectrum. *Psychological Medicine, 48*(2), 229–244. doi:10.1017/S0033291717001775 PMID:28689498

Gupta, S., Khandelwal, S.K., Tandon, P.N., Maheshwari, M.C., Mehta, V.S., Sundram, K.R., … Jain, S. (2000). The Development and Standardization of a Comprehensive Neuropsychological Battery in Hindi -Adult form. *Journal of Personality and Clinical Studies, 16*(2), 75–109.

Halpern, M. E., Gunturkun, O., Hopkins, W. D., & Rogers, L. J. (2005). Lateralization of the Vertebrate Brain: Taking the Side of Model Systems. *The Journal of Neuroscience: The Official Journal of the Society for Neuroscience, 25*(45), 10351–10357. doi:10.1523/JNEUROSCI.3439-05.2005 PMID:16280571

Hariri, A. R., & Weinberger, D. R. (2003). Imaging Genomics. *British Medical Bulletin, 65*(1), 259–270. doi:10.1093/bmb/65.1.259 PMID:12697630

Hart, S. J., Davenport, M. L., Hooper, S. R., & Belger, A. (2006). Visuospatial executive function in Turner syndrome: Functional MRI and neurocognitive findings. *Brain*, *129*(5), 1125–1136. doi:10.1093/brain/awl046 PMID:16504970

Healy, D. (2015). Serotonin and depression. *British Medical Journal*, ▪▪▪, 350.

Heaton, R. K., Chelune, G. J., Talley, J. L., Kay, G. G., & Curtiss, G. (1993). *Wisconsin Card Sorting Test Manual: Revised and Expanded*. Odessa, FL: Psychological Assessment Resources.

Hecht, D. (2010). Depression and the hyperactive right-hemisphere. *Neuroscience Research*, *68*(2), 77–87. doi:10.1016/j.neures.2010.06.013 PMID:20603163

Hechtman, L. (2019). ADHD medication treatment and risk of psychosis. *The Lancet. Psychiatry*, *6*(8), 632–633. doi:10.1016/S2215-0366(19)30248-2 PMID:31221558

Hendelman, W. (2005). *Atlas of functional neuroanatomy*. CRC Press. Retrieved from http://pages.britishlibrary.net/phrenology/

Henrik, L. (2010). Introduction. In Forming the Mind: Essays on the Internal Senses and the Mind/Body Problem from Avicenna to the Medical Enlightenment. Springer Science+Business Media.

Hernandes, M. S., & Troncone, L. R. (2009). Glycine as a neurotransmitter in the forebrain: A short review. *Journal of Neural Transmission (Vienna, Austria)*, *116*(12), 1551–1560. doi:10.100700702-009-0326-6 PMID:19826900

Hodapp, R. M., Dankner, N. A., & Dykens, E. M. (2016). Behavioural phenotypes/genetic syndromes. *Psychiatric and Behavioural Disorders in Intellectual and Developmental Disabilities*, 196.

Hofmann, A. (1970). *The Discovery of LSD and Subsequent Investigations on Naturally Occurring Hallucinogens in Discoveries in Biological Psychiatry*. JB Lippincott Company.

Hunt, J. (1949). On the localization of the functions of the brain with special reference faculty of language (Part II). *Anthropological Review*, *7*(24), 100–116. doi:10.2307/3025013

Iordanova, R., & Reddivari, A. K. R. (2019). Neuroanatomy, Medulla Oblongata. StatPearls Publishing.

Jackson, T., & Ramaswami, M. (2003). Prospects of memory-modifying drugs that target the CREB pathway. *Current Opinion in Drug Discovery & Development*, *6*, 712–719. PMID:14579521

James, W. (1890). *The principles of psychology*. New York: Holt.

Jellinger, K. A. (2010). Should the word 'dementia' be forgotten? *Journal of Cellular and Molecular Medicine*, *14*(10), 2415–2416. doi:10.1111/j.1582-4934.2010.01159.x PMID:20731751

Jennett, B., & Teasdale, G. (1981). *Management of head injuries*. Philadelphia: F. A. Davis Company.

Jewett, B. E., & Sharma, S. (2019). *Physiology, GABA*. Treasure Island, FL: Stat Pearls Publishing.

John, J. P., & Pinel, P. (2011). *Biopsychology* (8th ed.). Academic Press.

Juni, J. E., Waxman, A. D., Devous, M. D., Tikofsy, R. S., Ichise, M., VanHeertum, R. L., & ... . (2009). Procedure guideline for brain perfusion SPECT Using 99mTc Radiopharmaceuticals 3.0. *Journal of Nuclear Medicine Technology*, *37*(3), 191–195. doi:10.2967/jnmt.109.067850 PMID:19692453

Kandel, E. R. (2007). *In search of memory: The emergence of a new science of mind.* W. W. Norton & Company.

Kandel, E., Schwartz, J., & Jessel, T. (2000). *Principles of Neural Science* (4th ed.). New York: McGraw–Hill.

Kaplan, E. (1988). The process approach to neuropsychological assessment. *Aphasiology*, *2*(3–4), 309–311. doi:10.1080/02687038808248930 PMID:2136065

Kapur, S., Craik, F. I. M., Tulving, E., Wilson, A. A., Houle, S., & Brown, G. M. (1994). Neuroanatomical correlates of encoding in episodic memory: Levels of processing effect. *Proceedings of the National Academy of Sciences of the United States of America*, *91*(6), 2008–2011. doi:10.1073/pnas.91.6.2008 PMID:8134340

Kapur, S., Houle, S., & Brown, M. G. (1994). Positron Emission Tomographhy in Psychiatry: New Sights, New Insights. *Jefferson Journal of Psychiatry*, *12*(1), 58–74. doi:10.29046/JJP.012.1.007

Kar, B.R., Rao, S.L., Chandramouli, B.A., & Thennrasu, K. (2004). *NIMHANS Neuropsychological Battery for Children- Manual.* National Institute of Mental Health and Neurosciences.

Karaca, S., Saleh, A., Canan, F., & Potenza, M. N. (2017). Comorbidity between behavioral addictions and attention deficit/hyperactivity disorder: A systematic review. *International Journal of Mental Health and Addiction*, *15*(3), 701–724. doi:10.100711469-016-9660-8

Karunamuni, N. D. (2015). The Five-Aggregate Model of the Mind. *SAGE Open*, *5*(2). doi:10.1177/2158244015583860

Kellermann, T. S., Bonilha, L., Eskandari, R., Garcia-Ramos, C., Lin, J. J., & Hermann, B. P. (2016). Mapping the neuropsychological profile of temporal lobe epilepsy using cognitive network topology and graph theory. *Epilepsy & Behavior*, *63*, 9–16. doi:10.1016/j.yebeh.2016.07.030 PMID:27532489

Kertesz, A., & Benson, D. F. (1970). Neologistic jargon: A clinicopathological study. *Cortex*, *6*(4), 362–386. doi:10.1016/S0010-9452(70)80002-8 PMID:5514511

Kessels, R. P. C., & Hendriks, M. P. H. (2016). Neuropsychological Assessment. In Encyclopedia of Mental Health (2nd ed.; vol. 3). Waltham, MA: Academic Press. doi:10.1016/B978-0-12-397045-9.00136-1

Kessels, R. P. C., & Brands, A. M. A. (2009). Neuropsychological assessment. In: Biessels, Larrabee, G.J. (2008). Flexible vs. fixed batteries in forensic neuropsychological assessment: Reply to Bigler and Hom. *Archives of Clinical Neuropsychology*, *23*(7–8), 763–776.

Kettenmann, H., & Verkhratsky, A. (2016). Advances of Experimental Medicine Biology. *Springer Nature (Switzerland)*, *949*, 1–24.

Kim, J. (1995). Emergent properties. In T. Honderich (Ed.), *Problems in the Philosophy of Mind. Oxford Companion to Philosophy* (p. 240). Oxford, UK: Oxford University Press.

Kincaid, D. L., Doris, M., Shannon, C., & Mulholland, C. (2017). What is the prevalence of autism spectrum disorder and ASD traits in psychosis? A systematic review. *Psychiatry Research*, *250*, 99–105. doi:10.1016/j.psychres.2017.01.017 PMID:28152400

Kircher, T., Schneider, F., Sauer, H., & Buchkremer, G. (2004). Funktionelle Bildgebung am Beispiel der Schizophrenie. *Deutsches Arzteblatt*, *101*, A–1975.

Knight, D. (1914). *An outline of psychobiology*. John Hopkins Press.

Kodirov, S. A., Takizawa, S., Joseph, J., Kandel, E. R., Shumyatsky, G. P., & Bolshakov, V. Y. (2006). Synaptically released zinc gates long-term potentiation in fear conditioning pathways. *Proceedings of the National Academy of Sciences of the United States of America*, *103*(41), 15218–15223. doi:10.1073/pnas.0607131103 PMID:17005717

Kolb, B., & Whishaw, I. Q. (2009). *Fundamentals of human neuropsychology*. New York, NY: Worth Publishers.

Kosten, T. R., Haile, C. N., Newton, T. F., & De La Garza II, R. (2018). Alcohol and Other Substance Use Disorders. *The American Psychiatric Association Publishing Textbook of Neuropsychiatry and Clinical Neurosciences*, 413.

Kotzbauer, P. T., Trojanowsk, J. Q., & Lee, V. M. (2001). Lewy body pathology in Alzheimer's disease. *Journal of Molecule Neuroscience, 17*(2), 225-32.

Kuffler, S. W., & Nicholls, J. G. (1976). *From Neuron to Brain*. Sunderland, MA: Sinauer Associates.

Kumar, S., Singh, R.K., & Meena, R. (2016). Emerging targets for radioprotection and radiosensitization in radiotherapy. *Tumor Biology, 37*(9), 11589-609.

Laruelle, M. (2000). Imaging synaptic neurotransmission with in vivo binding competition techniques: A critical review. *Journal of Cerebral Blood Flow and Metabolism*, *20*(3), 423–451. doi:10.1097/00004647-200003000-00001 PMID:10724107

Laruelle, M., Kegeles, L. S., & Abi-Dargham, A. (2003). Glutamate, dopamine, and schizophrenia: From pathophysiology to treatment. *Annals of the New York Academy of Sciences*, *1003*(1), 138–158. doi:10.1196/annals.1300.063 PMID:14684442

Leon-Carrion. (1986). *Diagnostico Clinico en Psicologia*. Sevilla: Alfar.

Lezak, M. D. (1976). *Neuropsychological assessment*. Oxford, UK: Oxford University Press.

Li, Y., Sun., H., Chen, Z., Xu, H., Bu, G., & Zheng, H. (2016). Implications of GABAergic Neurotransmission in Alzheimer's Disease. *Frontiers in Aging Neuroscience, 23*(8), 31. doi:10.3389/fnagi.2016.00031

Linardon, J., Phillipou, A., Newton, R., Fuller-Tyszkiewicz, M., Jenkins, Z., Cistullo, L. L., & Castle, D. (2018). Testing the relative associations of different components of dietary restraint on psychological functioning in anorexia nervosa and bulimia nervosa. *Appetite, 128*, 1–6. doi:10.1016/j.appet.2018.05.138 PMID:29807122

Linden, D. J. (2012). The challenges and promises of neuroimaging in Psychiatry. *Neuron, 73*(1), 8–22. doi:10.1016/j.neuron.2011.12.014 PMID:22243743

Linden, D. J., & Dierks, T. (2002). Funktionelle Bildgebung akustischer Halluzinationen bei der Schizophrenie. *Nervenheilkunde: Zeitschrift für Interdisziplinaere Fortbildung, 21*, 331–335.

Liu, C. N., & Chambers, W. W. (1958). Intraspinal sprouting of dorsal root axons. *Archives of Neurology, 79*, 46–61. PMID:13486981

Liz, S. (2016). *Dementia Rates Decline Sharply Among Senior Citizens*. Retrieved from https://khn.org/news/dementia-rates-decline-sharply-among-senior-citizens/

Lodish, H., Berk, A., Zipursky, S. L., Matsudaira, P., Baltimore, D., & Darnell, J. W. H. (2002). Molecular Cell Biology. Science.

Lopes, M. B. S. (2019). Diagnostic histochemistry in neuropathology. *Seminars in Diagnostic Pathology, 36*(1), 71–82. doi:10.1053/j.semdp.2018.10.004 PMID:30471893

Luchsinger, J.A. (Eds.). (2010). Diabetes and the Brain. Totowa, NJ: Humana Press.

Luria, A. R. (1966). *Higher cortical functions in man*. New York: Springer.

Mann, L. (1979). *On the trail of process: A historical perspective on cognitive processes and their training*. New York: Grune & Stratton.

Marx, K. (1977). A Contribution to the Critique of Political Economy. Progress Publishers.

Marx, K. (1977). *A Contribution to the Critique of Political Economy*. Moscow: Progress Publishers.

Masdeu, J. C. (2011). Neuroimaging in Psychiatric Disorders. Neurotherapeutics. *The Journal of the American Society for Experimantal Neurotherapeutics., 8*(1), 93–102. doi:10.100713311-010-0006-0 PMID:21274689

Masdeu, J. C., Zubieta, J. L., & Arbizu, J. (2005). Neuroimaging as a marker of the onset and progression of Alzheimer's disease. *Journal of the Neurological Sciences, 236*(1-2), 55–64. doi:10.1016/j.jns.2005.05.001 PMID:15961110

McAndrews, M. P., & Cohn, M. (2012). Neuropsychology in temporal lobe epilepsy: Influences from cognitive neuroscience and functional neuroimaging. *Epilepsy Research and Treatment, 2012*, 1–13. doi:10.1155/2012/925238 PMID:22957249

McGilchrist, I., & Shankar, V. (2019). One Head, Two Brains: How The Brain's Hemispheres Shape The World We See (Audio podcast with transcript). *NPR Hidden Brain*.

McHenry, L. C. (1969). *Garrioson's History of Neurology*. Springfield, IL: Charles C. Thomas.

Medical Dictionary. (2020). Retrieved from https://medical-dictionary.thefreedictionary.com/regeneration

Mesulam, M. (2004). The cholinergic lesion of Alzheimer's disease: Pivotal factor or side show? *Learning & Memory*, *11*(1), 43–49. doi:10.1101/lm.69204 PMID:14747516

Miller, G. M. (2011). The emerging role of trace amine-associated receptor 1 in the functional regulation of monoamine transporters and dopaminergic activity. *Journal of Neurochemistry*, *116*(2), 164–176. doi:10.1111/j.1471-4159.2010.07109.x PMID:21073468

Misra, K., & Pandey, S. C. (2003). Differences in basal levels of CREB and NPY in nucleus accumbens regions between C57BL/6 and DBA/2 mice differing in inborn alcohol drinking behavior. *Journal of Neuroscience Research*, *74*(6), 967–975. doi:10.1002/jnr.10831 PMID:14648603

Moise, D., & Madhusoodanan, S. (2006). Psychiatric symptoms associated with brain tumors: A clinical enigma. *CNS Spectrums*, *11*(1), 28–31. doi:10.1017/S1092852900024135 PMID:16400253

Moore, P. M., & Baker, G. A. (2002). The neuropsychological and emotional consequences of living with intractable temporal lobe epilepsy: Implications for clinical management. *Seizure*, *11*(4), 224–230. doi:10.1053eiz.2001.0668 PMID:12027568

Morgan, C., & Anderson, G. C. (2016). Childhood adversities and psychosis: Evidence, challenges, implications. *World Psychiatry; Official Journal of the World Psychiatric Association (WPA)*, *15*(2), 93–102. doi:10.1002/wps.20330 PMID:27265690

Nasreddine, Z. S., Phillips, N. A., Bédirian, V., Charbonneau, S., Whitehead, V., Collin, I., & Chertkow, H. (2005). The Montreal Cognitive Assessment, MoCA: A brief screening tool for mild cognitive impairment. *Journal of the American Geriatrics Society*, *53*(4), 695–699. doi:10.1111/j.1532-5415.2005.53221.x PMID:15817019

Netter, P. (1991). Biochemical variables in the study of temperament. In J. Strelau & A. Angleitner (Eds.), *Explorations in temperament: International perspectives on theory and measurement 147–161*. New York: Plenum Press. doi:10.1007/978-1-4899-0643-4_10

Newton & Johnson. (1985). Principles of Neuropsychology (2nd ed.). Academic Press.

NIMH. (2014). *Brain Basics*. U.S. National Institutes of Health.

Nockleby, D. M., & Deaton, A. V. (1987). Denial versus distress: Coping patterns in post head trauma patients. *The International Journal of Clinical Neuropsychology*, *9*(4), 145–148.

Nuss, P. (2015). Anxiety disorders and GABA neurotransmission: A disturbance of modulation. *Neuropsychiatric Disease and Treatment*, *11*, 165–175. PMID:25653526

Olds, J., & Milner, P. (1954). Positive reinforcement produced by electrical stimulation of septal area and other regions of rat brain. *Journal of Comparative and Physiological Psychology*, *47*(6), 419–427. doi:10.1037/h0058775 PMID:13233369

Orrison, W.W., Lewine, J., Sanders, J., & Hartshorne, M.F. (2017). Functional brain imaging. *Elsevier Health Sciences, 24.*

Pacula, R. L., Powell, D., Heaton, P., & Sevigny, E. L. (2015). Assessing the effects of medical marijuana laws on marijuana use: The devil is in the details. *Journal of Policy Analysis and Management, 34*(1), 7–31. doi:10.1002/pam.21804 PMID:25558490

Pagulayan, K. F., Hoffman, J. M., Temkin, N. R., Machamer, J. E., & Dikmen, S. S. (2008). Functional Limitations and Depression After Traumatic Brain Injury: Examination of the Temporal Relationship. *Archives of Physical Medicine and Rehabilitation, 89*(10), 1887–1892. doi:10.1016/j.apmr.2008.03.019 PMID:18929017

Pampallona, S., Bollini, P., Tibaldi, G., Kupelnick, B., & Munizza, C. (2004). Combined pharmacotherapy and psychological treatment for depression: a systematic review. *Archives of General Psychiatry, 61*(7), 714-9.

Pearl, P. L., & Gibson, K. M. (2004). Clinical aspects of the disorders of GABA metabolism in children. *Current Opinion in Neurology, 17*(2), 107–113. doi:10.1097/00019052-200404000-00005 PMID:15021235

Pershad, D., & Verma, S. K. (2007). Hand Book of PGI Battery of Brain Dysfunction (PGIBBD) (2nd ed.). National Psychological Corporation.

Pfrieger, F. W. (2010). Role of glial cells in the formation and maintenance of synapses. *Brain Research. Brain Research Reviews, 63*(1-2), 39–46. doi:10.1016/j.brainresrev.2009.11.002 PMID:19931561

Pinel, J. P. J. (2003). Biopsychology (5th ed.). Allyn and Bacon.

Pinel, J. (2009). *Psychobiology* (7th ed.). Pearson/Allyn and Bacon.

Prigatano, G. P. (1999). *Principles of Neuropsychological Rehabilitation*. Oxford University Press.

Prigatano, G. P. (2010). *Advances in the study of anosognosia*. New York: Oxford University Press.

Prigatano, G. P., & Leathem, J. M. (1993). Awareness of behavioral limitations after traumatic brain injury: A cross-cultural study of New Zealand Maoris and non-Maoris. *Clinical Neuropsychologist, 7*(2), 123–135. doi:10.1080/13854049308401514 PMID:29022474

Purisch, A. D. (2001). Misconceptions about the Luria-Nebraska Neuropsychological Battery. *NeuroRehabilitation, 16*(4), 275–280. doi:10.3233/NRE-2001-16412 PMID:11790914

Rao, S.L., Subbakrishna, D.K., & Gopukumar, K. (2004). *NIMHANS Neuropsychological Battery-Manual*. National Institute of Mental Health and Neurosciences.

Raven, P. H., & Johnson, G. B. (1992). *Biology* (3rd ed.). St. Louis, MO: Mosby Year Book.

Reinhard, S. C., Given, B., Petlick, N. H., & Bemis, A. (2008). Supporting Family Caregivers in Providing Care. In *Patient Safety and Quality: An Evidence-Based Handbook for Nurses*. Rockville, MD: Agency for Healthcare Research and Quality. Available from: https://www.ncbi.nlm.nih.gov/books/NBK2665/

Reitan, R. M. (1994). Ward Halstead's contributions to neuropsychology and the Halstead-Reitan Neuropsychological Test Battery. *Journal of Clinical Psychology*, *50*(1), 47–70. doi:10.1002/1097-4679(199401)50:1<47::AID-JCLP2270500106>3.0.CO;2-X PMID:8150995

Reitan, R. M., & Davidson, L. D. (1974). *Clinical neuropsychology: Current status and applications*. Washington, DC: Winston.

Richard, A. G., & Brian, C. (2012). Following Teaching of Psychology Into Middle Age: Changes, Contributors, and Contents. *Teaching of Psychology*, *40*(1), 6–14.

Ries, S. K., Dronkers, N. F., & Knight, R. T. (2016). Choosing words: Left hemisphere, right hemisphere, or both? Perspective on the lateralization of word retrieval. *Annals of the New York Academy of Sciences*, *1369*(1), 111–131. doi:10.1111/nyas.12993 PMID:26766393

Robert, M. Y. (1996). The mind–body problem. In Companion to the History of Modern Science. Taylor and Francis.

Robinson, H. (2011). Dualism. The Stanford Encyclopedia of Philosophy.

Robinson, T. E., & Becker, J. B. (1986). Enduring changes in brain and behavior produced by chronic amphetamine administration: A review and evaluation of animal models of amphetamine psychosis. *Brain Research*, *396*(2), 157–198. doi:10.1016/0165-0173(86)90002-0 PMID:3527341

Ropper, A. H., Adams, R. D., Victor, M., & Brown, R. H. (2005). Adams and Victor's principles of neurology (8th ed.). New York: McGraw-Hill Medical Pub. Division.

Rose, M., & Devine, J. (2014). Assessment of patient-reported symptoms of anxiety. *Dialogues in Clinical Neuroscience*, *16*(2), 197–211. PMID:25152658

Rowe, C. C., & Ackermann, U. (2007). Imaging beta-amyloid burden in aging and dementia. *Neurology*, *68*(20), 1718–172. doi:10.1212/01.wnl.0000261919.22630.ea PMID:17502554

Russell, S., & Norvig, P. (2010). *Artificial Intelligence: A Modern Approach* (3rd ed.). Prentice Hall.

Sadock, B. J., Kaplan, H. I., & Sadock, V. A. (2005). *Kaplan & Sadock's Comprehensive Textbook of Psychiatry*. Philadelphia: Lippincott Williams & Wilkins.

Saling, M. M. (2009). Verbal memory in mesial temporal lobe epilepsy: Beyond material specificity. *Brain*, *132*(3), 570–582. doi:10.1093/brain/awp012 PMID:19251757

Santoro, J., & Spiers, M. (1994). Social cognitive factors in brain injury-associated personality change. *Brain Injury: [BI]*, *8*(3), 265–276. doi:10.3109/02699059409150978 PMID:8004084

Santra, A., & Kumar, R. (2014). Brain perfusion single photon emission computed tomography in major psychiatric disorders: From basics to clinical practice. *Indian Journal of Nuclear Medicine*, *29*(4), 210–221. doi:10.4103/0972-3919.142622 PMID:25400359

Sarter, M. F., & Bruno, J. P. (1994). Cognitive functions of cortical ACh: Lessons from studies on trans-synaptic modulation of activated efflux. *Trends in Neurosciences*, *17*(6), 217–221. doi:10.1016/0166-2236(94)90001-9 PMID:7521080

Sarter, M., Bruno, J., & Parikh, V. (2007). Abnormal Neurotransmitter Release Underlying Behavioral and Cognitive Disorders: Toward Concepts of Dynamic and Function-Specific Dysregulation. *Neuropsychopharmacol*, *32*(7), 1452–1461. doi:10.1038j.npp.1301285 PMID:17164812

Sarter, M., & Parikh, V. (2005). Choline transporters, cholinergic transmission and cognition. *Nature Reviews. Neuroscience*, *6*(1), 48–56. doi:10.1038/nrn1588 PMID:15611726

Sbordone, R. J., Ronald, E., & Purisch, D. (2007). *Neuropsychology for Psychologists, Health Care Professionals and Attorneys*. CRC Press.

Schacter, D.L., Gilbert, D.T., & Wegner, D.M. (2009). *Psychology*. Academic Press.

Sherer, M., Boake, C., Levin, E., Silver, B.V., Ringholz, G., & Walter, M.H. (1998). Characteristics of impaired awareness after traumatic brain injury. *Journal of the International Neuropsychology Society, 4*, 380-387.

Sherer, M., Hart, T., & Todd, N. (2003). Measurement of impaired self-awareness after traumatic brain injury: A comparison of the patient competency rating scale and the awareness questionnaire. *Brain Injury: [BI]*, *17*(1), 25–37. doi:10.1080/0269905021000010113 PMID:12519645

Sherwin, I., Peron-Magnan, P., Bancaud, J., Bonis, A., & Talairach, J. (1982). Prevalence of psychosis in epilepsy as a function of the laterality of the epileptogenic lesion. *Archives of Neurology*, *39*(10), 621–625. doi:10.1001/archneur.1982.00510220019004 PMID:6812552

Skirry, J. (2016). *Rene Descartes: The Mind-Body Distinction*. Internet Encyclopedia of Philosophy.

Snyder, S. H., & Innis, R. B. (1979). Peptide neurotransmitters. *Annual Review of Biochemistry*, *48*(1), 755–782. doi:10.1146/annurev.bi.48.070179.003543 PMID:38738

Sohlberg, M. M., & Mateer, C. A. (1987). Effectiveness of an attention training program. *Journal of Clinical and Experimental Neuropsychology*, *19*(2), 117–130. doi:10.1080/01688638708405352 PMID:3558744

Sourkes, T. L. (1961). The biochemistry of mental disease. *Canadian Medical Association Journal*, *85*(9), 487. PMID:20326876

Spiers, M. V., Pouk, J. A., & Santoro, J. M. (1994). Examining perspective-taking in the severely head-injured. *Brain Injury: [BI]*, *8*(5), 463–473. doi:10.3109/02699059409150998 PMID:7951209

Squire, L. R. (2008). *Fundamental neuroscience* (3rd ed.). Academic Press.

Tai, Y. F., & Piccini, P. (2004). Applications of positron emission tomography (PET) in neurology. *Journal of Neurology, Neurosurgery, and Psychiatry*, *75*(5), 669–676. doi:10.1136/jnnp.2003.028175 PMID:15090557

Tamietto, M., & Leopold, D. A. (2018). Visual cortex: The eccentric area prostriata in the human brain. *Current Biology*, *28*(1), R17–R19. doi:10.1016/j.cub.2017.11.006 PMID:29316413

Taylor, I., & Taylor, M. M. (1990). *Psycholinguistics: Learning and using Language*. Pearson.

Terni, B., López-Murcia, F.J., & Llobet, A. (2011). Neuroglia living nerve glue. *Fortschrtte der Neurologie Psychiatrie, 79*(10), 588-97.

Terni, B., López-Murcia, F. J., & Llobet, A. (2016). Role of neuron-glia interactions in developmental synapse elimination. *Brain Research Bulletin*, *129*, 74–81. doi:10.1016/j.brainresbull.2016.08.017 PMID:27601093

Teufel, C., Ghazanfar, A. A., & Fischer, J. (2010). On the relationship between lateralized brain function and orienting asymmetries. *Behavioral Neuroscience*, *124*(4), 437–445. doi:10.1037/a0019925 PMID:20695643

The ICD-10. (1993). *Classification of Mental and Behavioural Disorders, Clinical descriptions and diagnostic guidelines*. World Health Organization.

Thomsen, I. V. (1974). The patient with severe head injury and his family. *Scandinavian Journal of Rehabilitation Medicine*, *6*, 180–183. PMID:4458060

Tomasi, D., & Volkow, N. D. (2019). Association between brain activation and functional connectivity. *Cerebral Cortex*, *29*(5), 1984–1996. doi:10.1093/cercor/bhy077 PMID:29668865

Trofimova, I. N. (2016). The interlocking between functional aspects of activities and a neurochemical model of adult temperament. In M. C. Arnold (Ed.), *Temperaments: Individual Differences, Social and Environmental Influences and Impact on Quality of Life* (pp. 77–147). New York: Nova Science Publishers, Inc.

Trofimova, I., & Robbins, T. W. (2016). Temperament and arousal systems: A new synthesis of differential psychology and functional neurochemistry. *Neuroscience and Biobehavioral Reviews*, *64*, 382–402. doi:10.1016/j.neubiorev.2016.03.008 PMID:26969100

Tyerman, A., & King, N. S. (2013). Interventions for psychological problems after brain injury. In L. H. Goldstein & J. E. McNeil (Eds.), *Clinical neuropsychology: A practical guide to assessment and management for clinicians* (pp. 527–548). Wiley-Blackwell.

Vaishali, H. P. (2019). Assessment of organicity in India. *International Journal of Research and Analytical Reviews, 6*(2), 322-26.

Vallar, G., & Perani, D. (1986). The anatomy of unilateral neglect after right-hemisphere stroke lesions. A clinical/CT-scan correlation study in man. *Neuropsychologia*, *24*(5), 609–622. doi:10.1016/0028-3932(86)90001-1 PMID:3785649

Van Praag, H. M. (1981). Management of depression with serotonin precursors. *Biological Psychiatry, 16*(3), 291–310. PMID:6164407

Vanderploeg, R. D. (Ed.). (2000). *Clinician's Guide to Neuropsychological Assessment* (2nd ed.). Hillsdale, NJ: Lawrence Erlbaum.

Vanderploeg, R. D., Curtiss, G., Luis, C. A., & Salazar, A. M. (2007). Long-term morbidities following self-reported mild traumatic brain injury. *Journal of Clinical and Experimental Neuropsychology, 29*(6), 585–598. doi:10.1080/13803390600826587 PMID:17691031

Verdon, V., Schwartz, S., Lovblad, K. O., Hauert, C. A., & Vuilleumier, P. (2010). Neuroanatomy of hemispatial neglect and its functional components: A study using voxel-based lesion-symptom mapping. *Brain, 133*(3), 880–894. doi:10.1093/brain/awp305 PMID:20028714

Vissers, L. E., Gilissen, C., & Veltman, J. A. (2016). Genetic studies in intellectual disability and related disorders. *Nature Reviews. Genetics, 17*(1), 9–18. doi:10.1038/nrg3999 PMID:26503795

von Bernhardi, R., Eugenín-von Bernhardi, J., Flores, B., & Eugenín León, J. (2016). Glial Cells and Integrity of the Nervous System. *Advances of Experimental Medicine, 949*, 1–24. doi:10.1007/978-3-319-40764-7_1 PMID:27714682

Vyas, J., & Ahuja, N. (2011). Textbook of Postgraduate Psychiatry (2nd ed..). Jaypee Brothers, Medical Publishers (p) Ltd.

Walsh, P., Kane, N., & Butler, S. (2005). The clinical role of evoked potentials. *Journal of Neurology, Neurosurgery, and Psychiatry, 76*(suppl_2), 16–22. doi:10.1136/jnnp.2005.068130 PMID:15961863

Waxman, A.D., Karl, H., David, H., Peter, H., Satoshi, M., & Masanori, I. (2009). *Procedure Guideline for FDG-PET Brain Imaging v1.01*. Society of Nuclear Medicine Procedure Guideline for FDG PET Brain Imaging Version 1.0.

West, R. L. (1995). Compensatory strategies for age associated memory impairment. In Hand Book of Memory Disorders (pp. 481-500). Wiley.

Whishaw, B. K., & Ian, Q. (2014). An introduction to brain and behavior (4th ed.). New York, NY: Worth Publishers.

Whooley, M. A., Grady, D., & Cauley, J. A. (2000). Postmenopausal estrogen therapy and depressive symptoms in older women. *Journal of General Internal Medicine, 15*(8), 535–541. doi:10.1046/j.1525-1497.2000.04029.x PMID:10940144

Wise, K., Ownsworth, T., & Fleming, J. (2005). Convergent validity of self-awareness measures and their association with employment outcome in adults following acquired brain injury. *Brain Injury: [BI], 19*(10), 765–775. doi:10.1080/0269905050019977 PMID:16175837

Yang, J. L., Sykora, P., Wilson, D. M. III, Mattson, M. P., & Bohr, V. A. (2011). The excitatory neurotransmitter glutamate stimulates DNA repair to increase neuronal resiliency. *Mechanisms of Ageing and Development, 132*(8-9), 405–411. doi:10.1016/j.mad.2011.06.005 PMID:21729715

Zald, D.H., Lee, J.T., Fluegel, K.W., & Pardo, J.V. (1998). Aversive gustatory stimulation activates limbic circuits in humans. *Brain: A Journal of Neurology, 121*(6), 1143-54.

Zillmer, E., Spiers, M., & Culbertson, W. C. (2008). Principles of neuropsychology (Recent Edition). Thomson/Wadsworth.

Zobel, A., Joe, A., Freymann, N., Clusmann, H., Schramm, J., Reinhardt, M., ... Broich, K. (2005). Changes in regional cerebral blood flow by therapeutic vagus nerve stimulation in depression: An exploratory approach. *Psychiatry Research, 139*(3), 165–179. doi:10.1016/j.pscychresns.2005.02.010 PMID:16043331

About the Contributors

Barre Vijay Prasad completed Master's Degree in Psychology from Andhra University, he later went to acquire his first M Phil in Rehabilitation Psychology from National Institute for the Mentally Handicapped (NIMH). Thereafter, he went on to acquire his second M Phil in Clinical Psychology from one of the most reputed institutes in India Institute of Human Behaviour and Allied Sciences (IHBAS), Delhi. He is the first psychologist to have accreditation as a dually trained professional psychologist with qualifications as a Clinical Psychologist and a Rehabilitation Psychologist from Rehabilitation Council of India (RCI) - Delhi. He is a Professional Life Member of Indian Association of Clinical Psychology (IACP). He has awarded Doctoral Degree in Psychology (Psychooncology) from Centre for Health Psychology, University of Hyderabad. He has been awarded: Centenary Merit award during beginning of his professional psychology career. Currently he is working as Assistant Professor of Clinical Psychology in department of Psychiatry at All India Institute of Medical Sciences (AIIMS)-New Delhi, prior to this Institute he worked as Associate Professor and Head of the Department of Clinical Psychology at Dharwad Institute of Mental Health and Neurosciences (DIMHANS-India). He is now working to produce ways of working manuals for reducing psychological distress amongst mentally ill and patients with cancer in the field of Psycho-oncology. He is trained at conducting various assessments such as Psycho-diagnostics, personality assessments amongst other skills. Not only does he work on assessments he is proficient in providing interventions using multi-modal approaches. He is trained in providing support for various initiatives disability rehabilitation, promotion of mental health and wellbeing. He specializes in cognitive behavioral therapies and behavioural interventions for various clinical disorders. He has published many articles in national and international journals, book chapters and books as well, and presented several conference papers on mental health, disability rehabilitation and psycho-oncology.

* * *

Shazia Ali is a Senior Resident at department of Advanced Centre for Human Gentics, Sher-i-Kashmir Institute of Medical Sciences (SKIMS). She completed her Ph.D Life Sciences, Breast Cancer Biology, Jawaharlal Nehru University, New Delhi. She is Technical Expertise in Biochemical techniques, SDS-PAGE, Western blotting. Molecular Diagnostic Techniques, Immuno blotting techniques, ELISA. Molecular Biology and Molecular genetics techniques, Microbial Biochemistry, Animal Tissue Culture Primary cell culture and routine cell culture techniques, handling animals, Special techniques learned through projects MTT -Cell-Viable assay, BrDu-Cell-Proliferation assay, Transfections, Cloning, iNOX assay-anti-inflammatory assay, DNA-laddering, TUNNEL-apoptosis assay, Caspase-assay, FACS etc., She is holding various reputed Membership in Scientific/ Professional Organizations and she is also editor of many journals and has published many national and international journals.

Manoj Kumar Bajaj is a licensed clinical psychologist. Major area of interest includes clinical neuropsychology, cognitive behavior therapy, Behavioral medicine. He is a Core faculty of M.Phil Clinical Psychology Program in GMCH India. He has been conducting workshops on Clinical neuropsychological assessments and interventions.

Kaloiya G. S. is currently working as Addl Professor of Clinical Psychology in Department of Psychiatry at NDDTC, AIIMS-India. His areas of Interest are Cognitive Behaviour Therapy (CBT), Group Therapy, Family Therapy, Addiction Psychology and OCD. Prior to working at AIIMS-NDDTC, he worked at various institutes as Consultant Clinical Psychologist. He supervised many PhD's and MD thesis. He published number of articles in national and International Journals.

Syed Irtiza he is working as Senior Resident at Human Genetic Department, SKIMS-India.

Ramprasad Santhanakrishnan K. is a professor and Head in Department of Psychiatry at Dharwad Institute of Mental Health and Neurosciences (DIMHANS), he did M.B.B.S from Katurba Medical College, Mangalore (Mangalore University), and FDAD from WHO recognized Fellowship in Drug and Alcohol de-addiction (New South Wales, Australia). Prior to DIMHANS, he worked as faculty at various institutes, his areas of research are Addiction Psychiatry, he published his research articles in national international journals.

Pradeep Kallollimath is currently working as Assistant Professor of Neurology at DIMHANS/KIMS, He did his MBBS from KIMS, Hubli (1998-2004); MD in General Medicine from VIMS, Bellary (2005-2008) and DM Neurology from NIMHANS, Bangalore (2010-2013). His areas of interest are stroke, epilepsy and Wilson's disease. He is a life member of Indian medical association, life member of Indian academy of neurology and member of American academy of neurology. He presented papers in national and international conferences. He has many publication and book chapters' national and international journals and books.

Harsimarpreet Kaur completed PhD in Neuropsychology from All India Institute of Medical Sciences (AIIMS), New Delhi. Presently working as an Assistant Professor, School of Human Ecology, Tata Institute of Social Sciences.

Sumit Kumar is completed Master's Degree in Biotechnology with neuroscience as a subject from The Maharaja Sayajirao University of Baroda, he later joined Madurai Kamaraj University, Madurai as Department of Biotechnology Junior Research Fellow. Thereafter, he went on to acquire his Doctor of Philosophy from School of Life Sciences, Jawaharlal Nehru University, New Delhi in memory consolidation during wakefulness and sleep in mice under Prof. Birendra Nath Mallick. At Jawaharlal Nehru University, he used single neuron recording system and maze to study area-specific neuronal firing during learning and memory consolidation in real-time. He has awarded Doctoral Degree in Life Sciences, Jawaharlal Nehru University, New Delhi. Currently, he is working as a Young Scientist at the Department of Biosciences, Jamia Millia Islamia, New Delhi. He has published many articles in peer-reviewed international journals, presented several conference papers on the above field. He is also working on writing a book on "Anticancer Therapeutics: past present and future".

Shahzadi Malhotra is a postgraduate in Psychology (2005) from University of Delhi obtained her professional degree M.Phil. in Clinical Psychology(University of Delh) from Institute of Human Behaviour and Allied Sciences in the year 2007. She completed her Ph.D from the prestigious Jamia Millia Islamia University, Delhi. As per her special area of interest in Special Education and Psychosocial Rehabilitation she thoroughly devoted herself to this area and obtained Post Graduate Diploma in Special Education from University of Delhi. She served reputed Special Education centres of Delhi in the capacity of Consultant Clinical Psychologist and also provided her services as Clinical Psychologist to the Department of Neurology and Department of Clinical Psychology at IHBAS Delhi. Currently (from the year 2010 onwards) she is serving Chacha Nehru Bal Chikitsalaya; a reputed hospital for children of Govt.of NCT of Delhi as faculty: Assistant Professor of Clinical Psychology. She has been awarded for her valuable contribution to the discipline by various professional associations including Indian Association of Clinical Psychologists, and Indian Psychiatric Society- North zone, Delhi Psychiatry Society and Indian Association of Cognitive Behaviour therapy.

Arpit Parmar is a DM (addiction psychiatry) resident from NDDTC and Psychiatry, All India Institute of Medical Sciences, New Delhi, India. He has completed MD Psychiatry from the same institute and has authored more than 40 research articles in the field of psychiatry and addiction medicine. His areas of interest include addiction and biological psychiatry.

Gaurav Rajender is a postgraduate in Psychiatry from Institute of Human Behaviour and Allied Sciences. He has served in various institutes of repute like Guru Teg Bahadur Hospital, G.B. Pant Hospital, All India Institute of Medical Sciences. Currently he is working as faculty of Psychiatry at SMS Medical college Jaipur. He has been awarded for his valuable contribution to the discipline by various professional associations including Indian Psychiatric Society-North zone, Indian Association of social Psychiatry, Delhi Psychiatry Society.

Index

G

H

I

K

L

M

N

O

P

Q

S

www.ingramcontent.com/pod-product-compliance
Lightning Source LLC
LaVergne TN
LVHW081522300926
849161LV00006B/1269